The Cattle Hustler

The Cattle Hustler

Colouring Book

Jay Elliott

ColorpopAI

J L Elliott

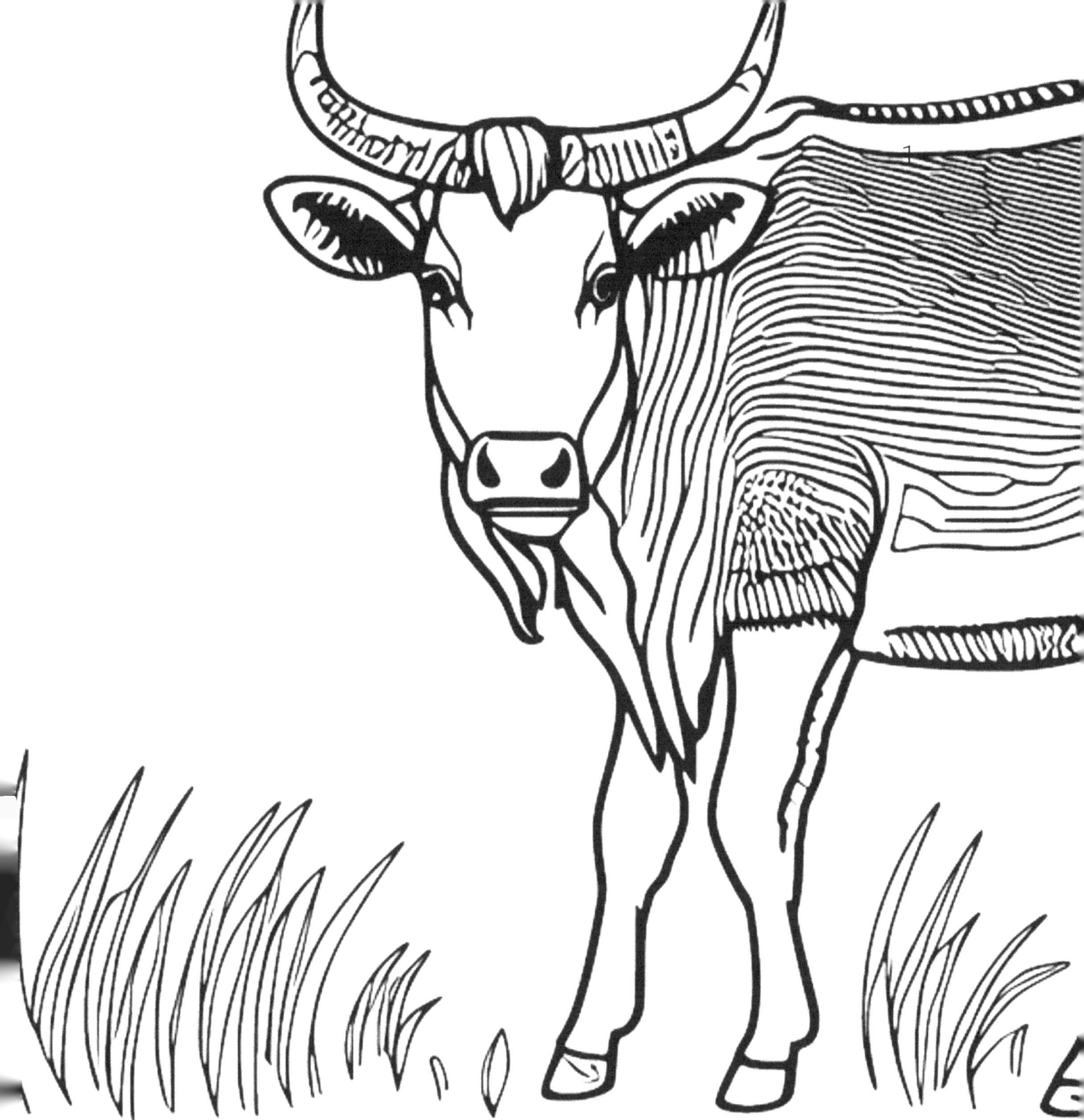

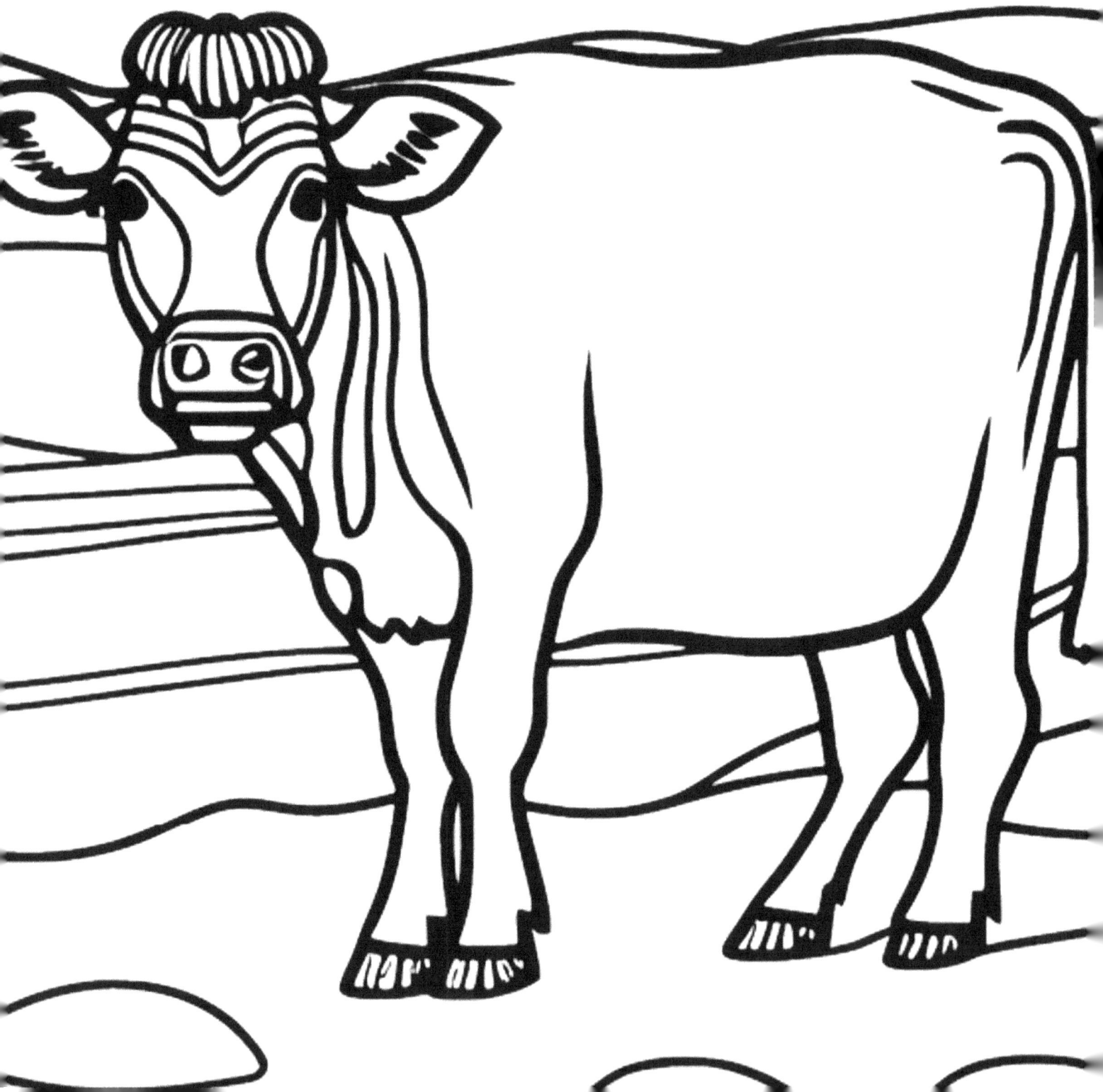

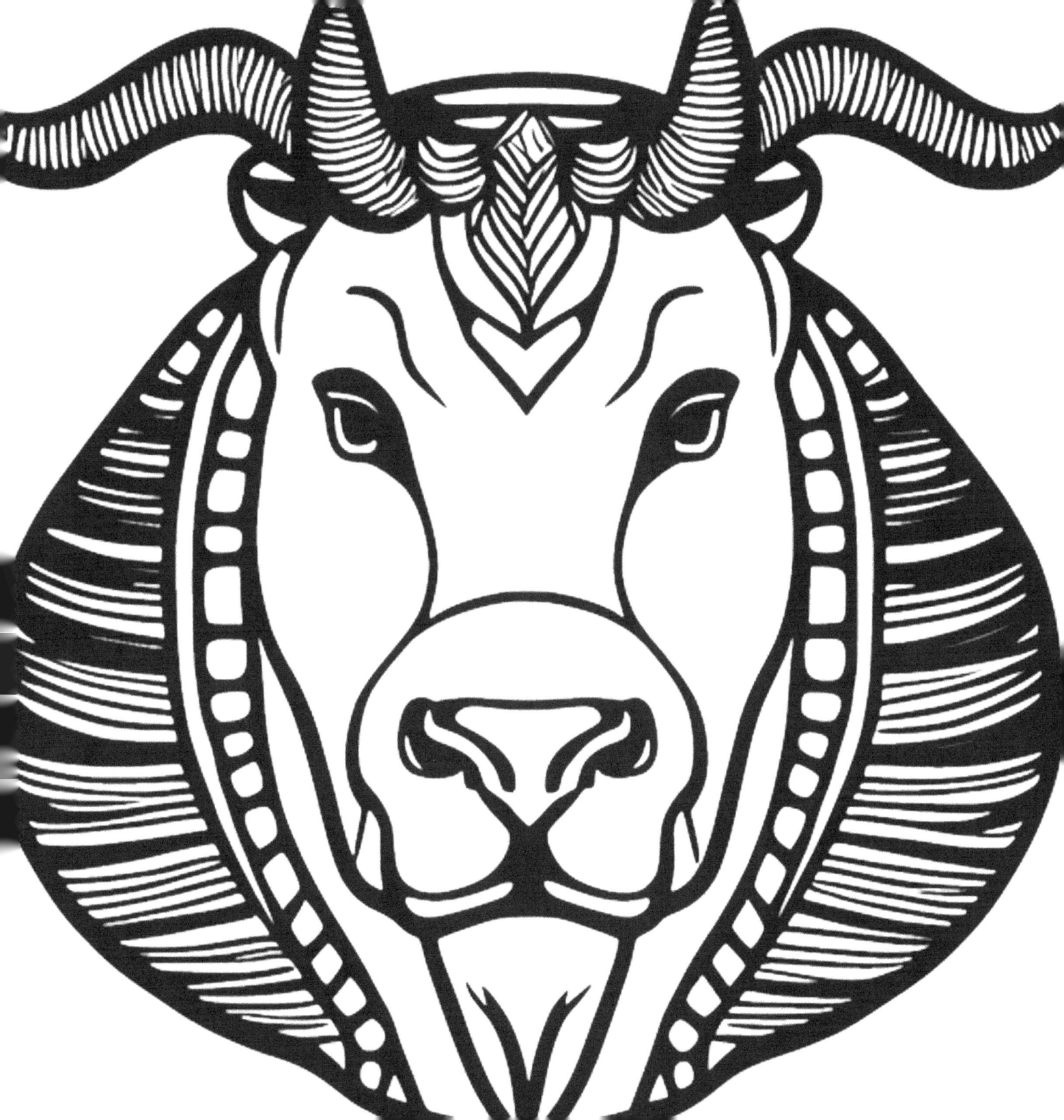

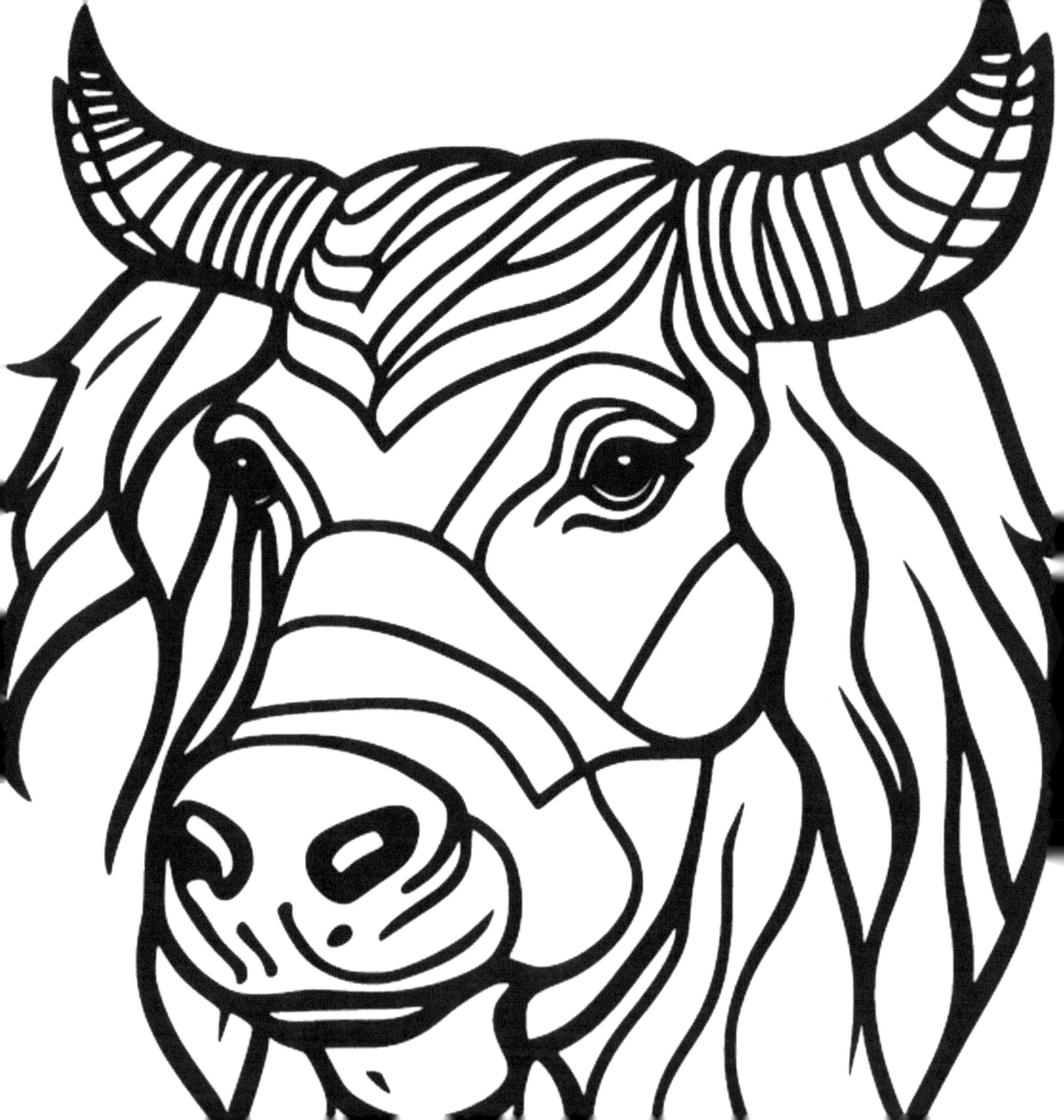

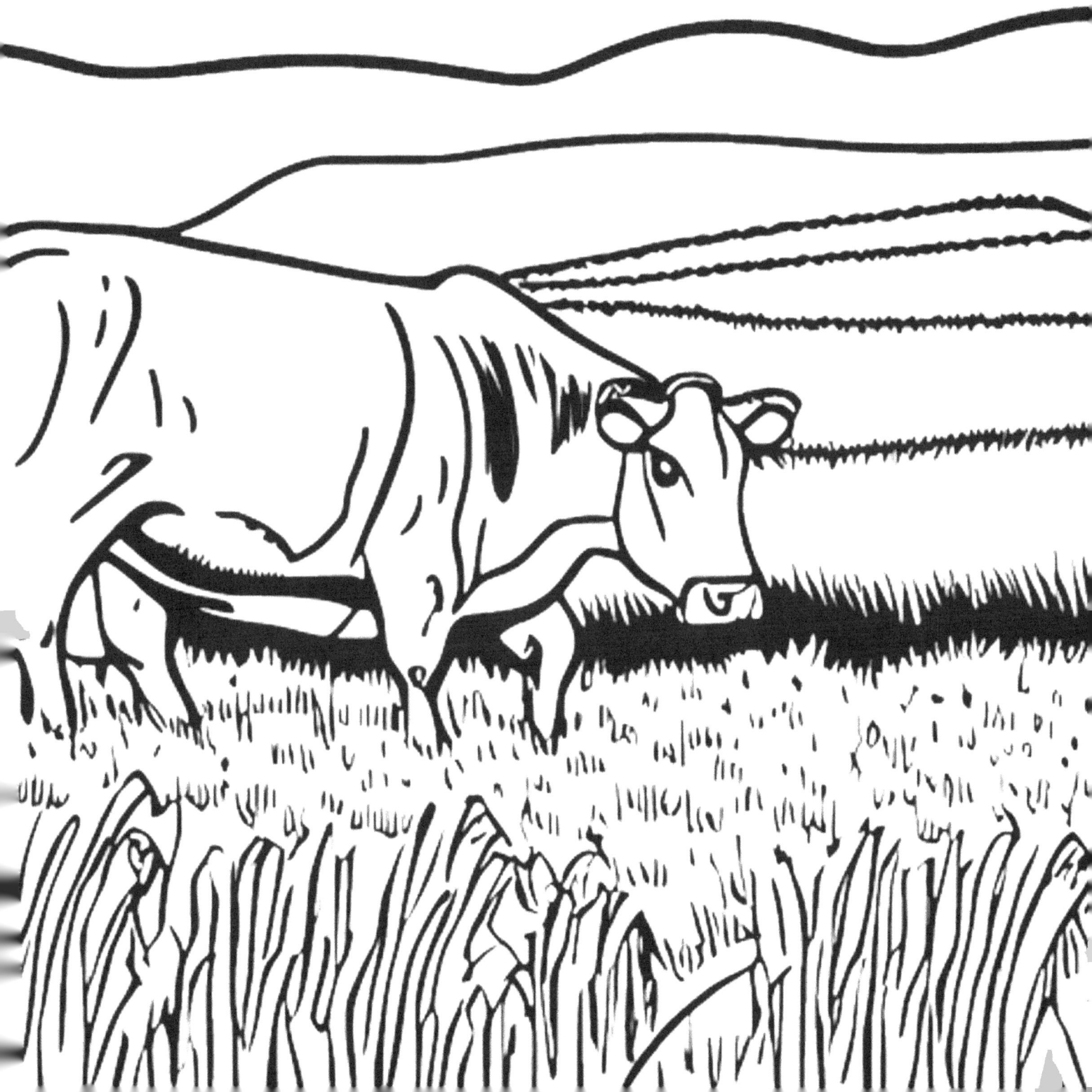

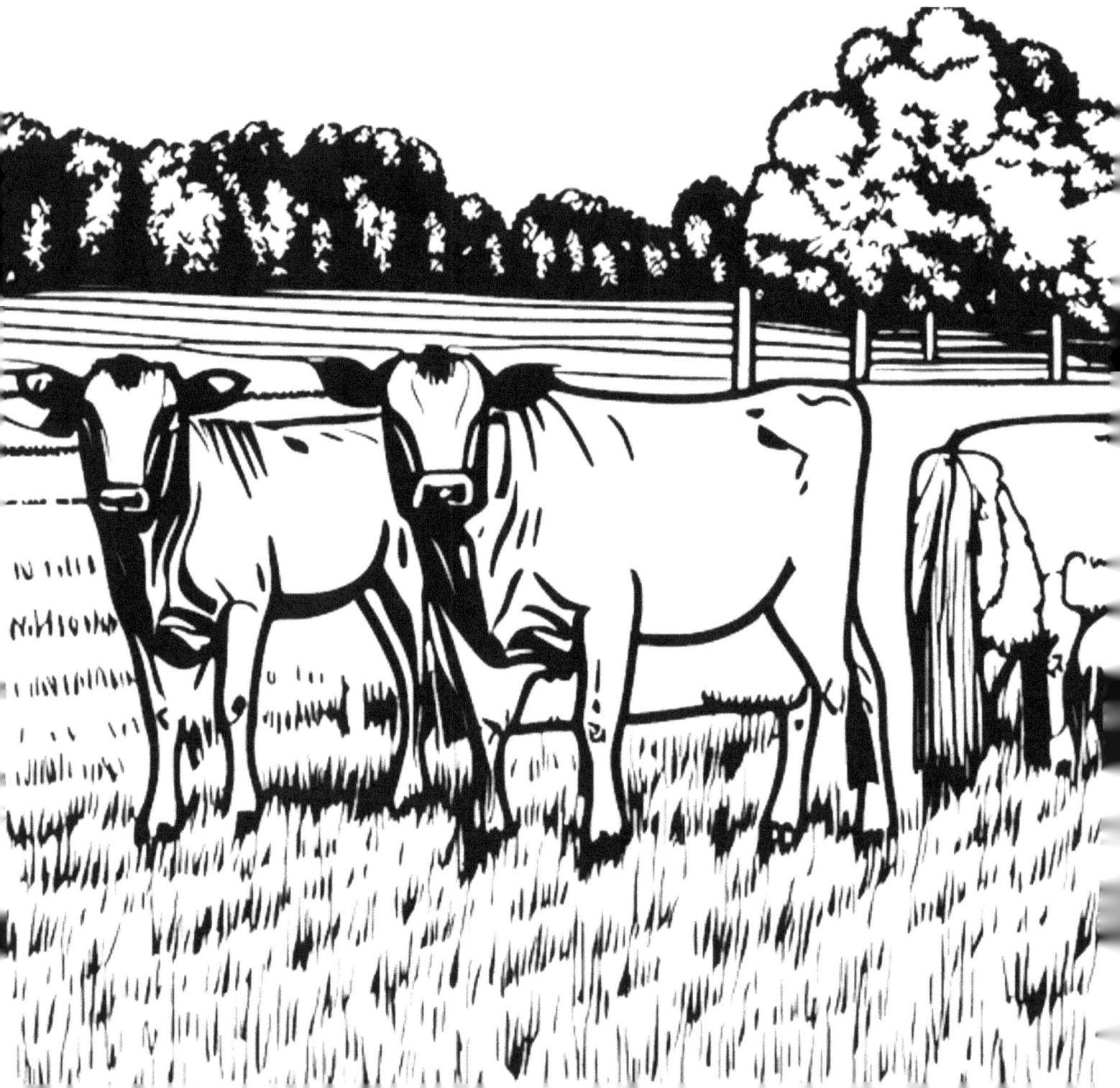

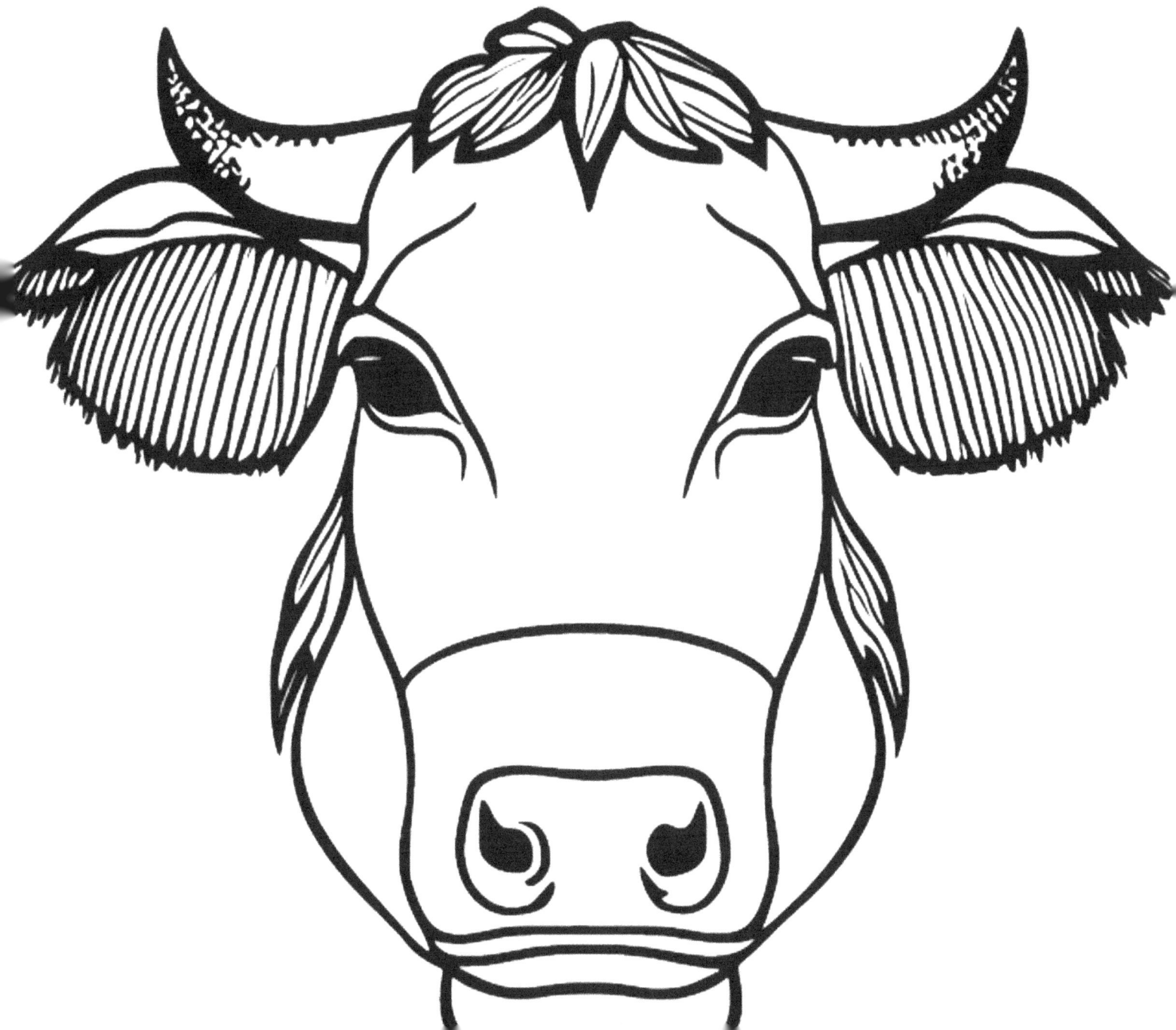

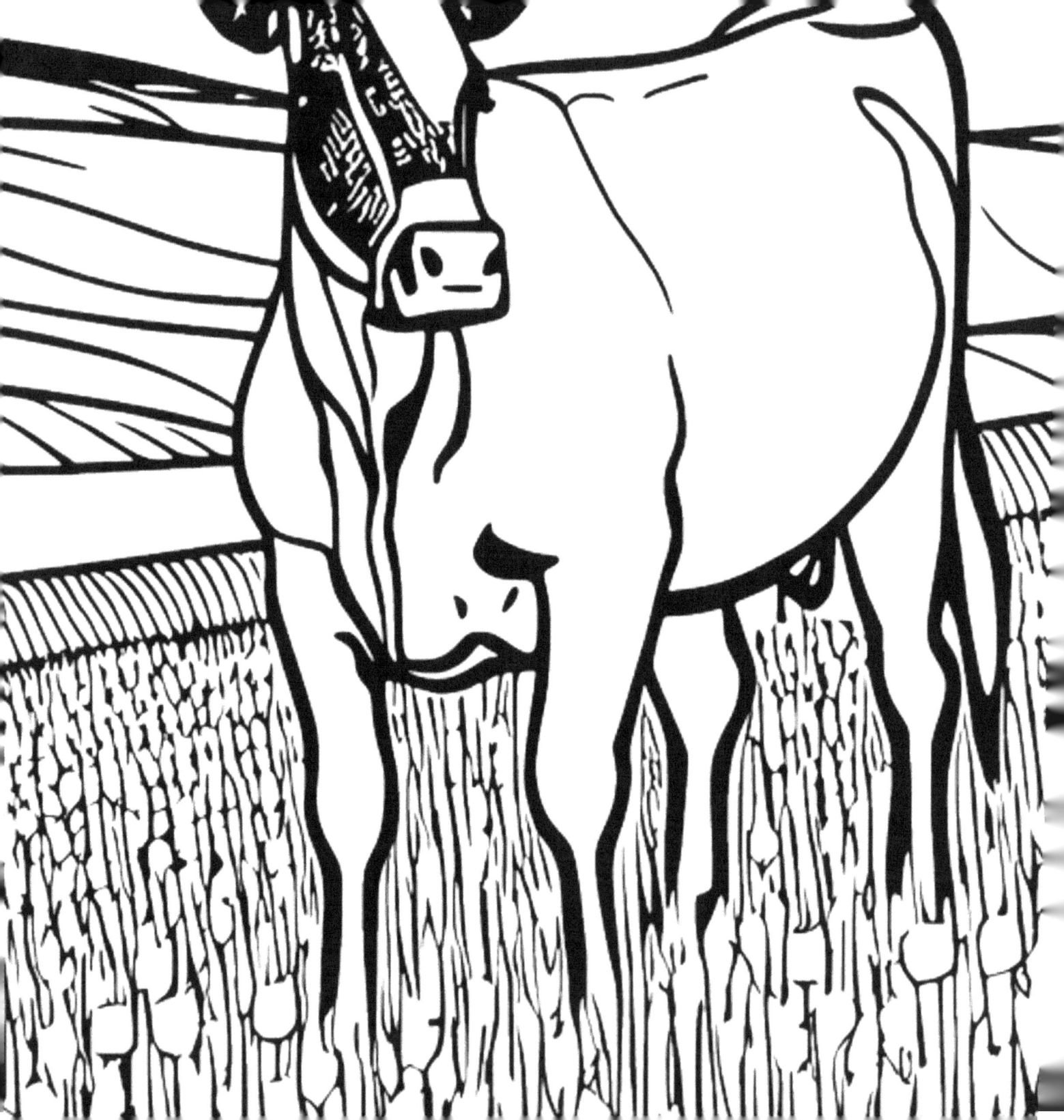

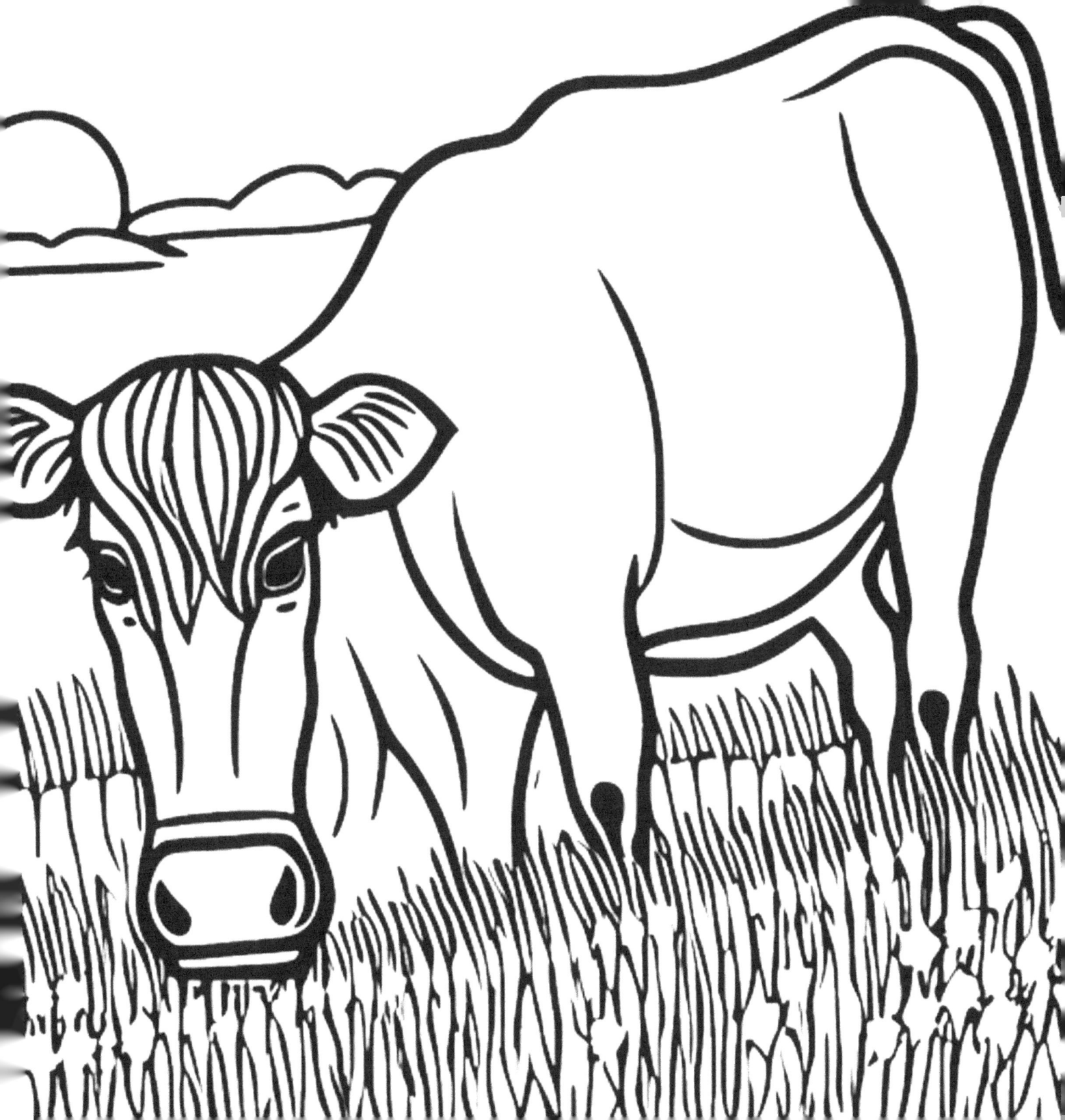

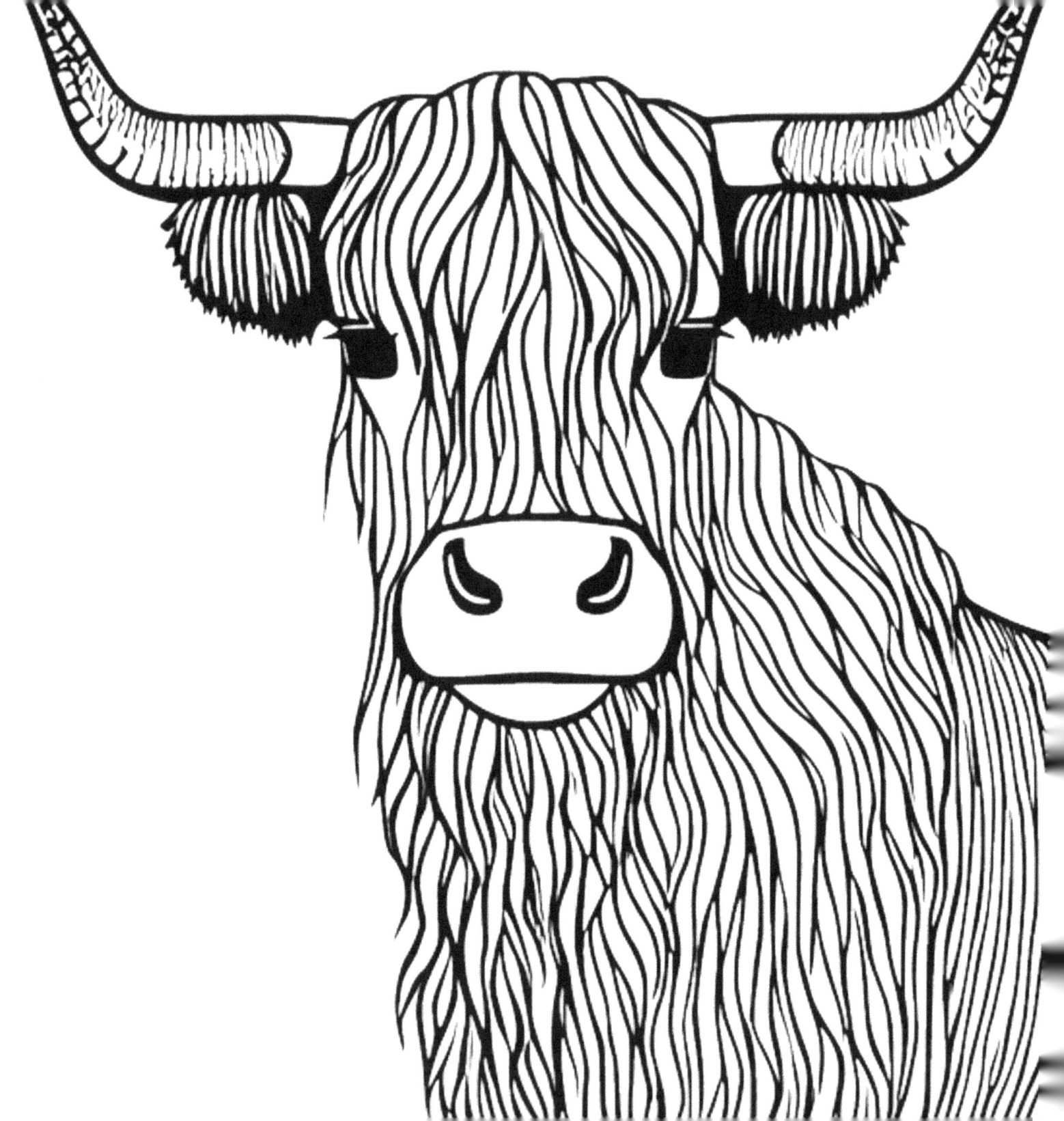

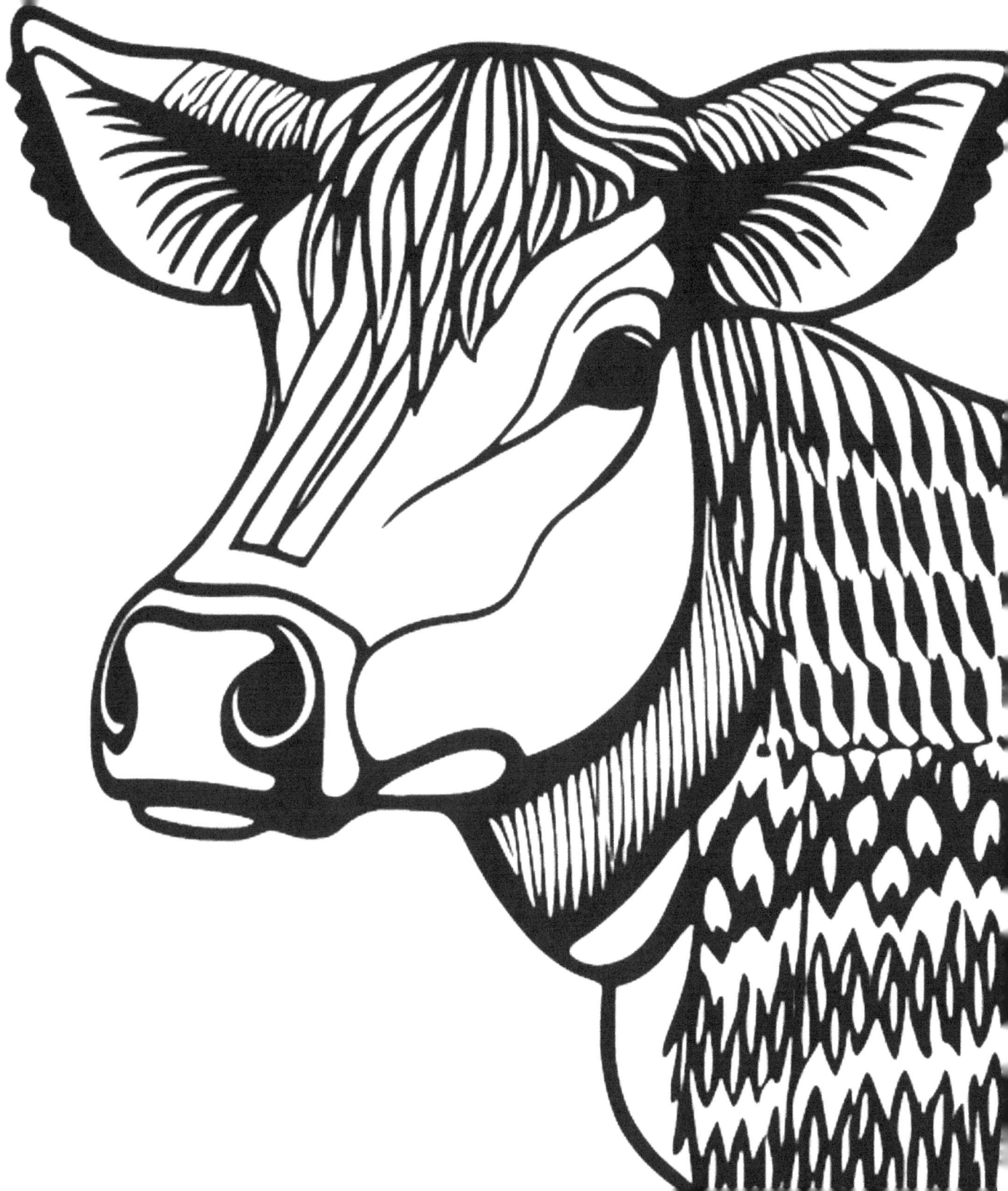

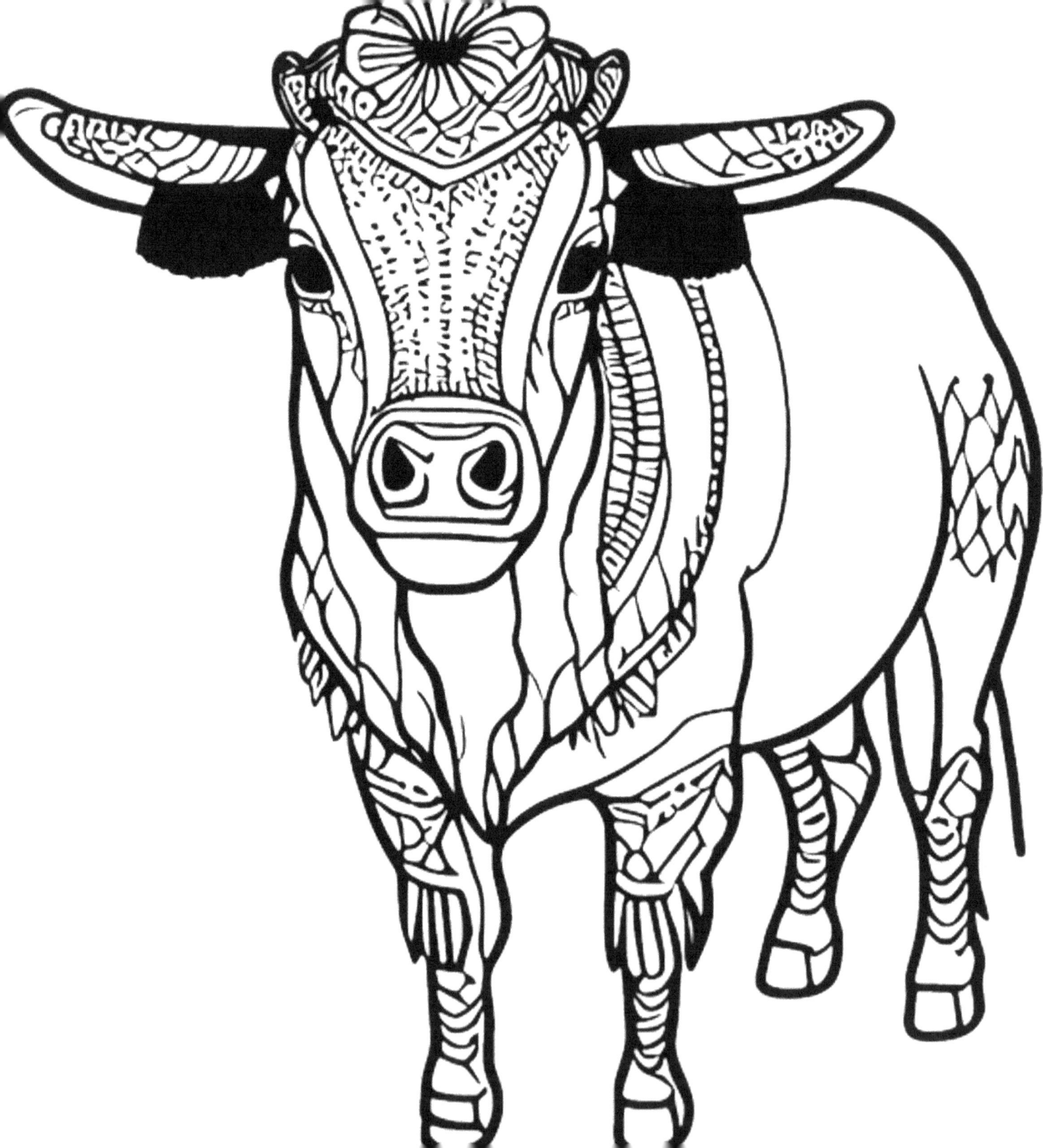

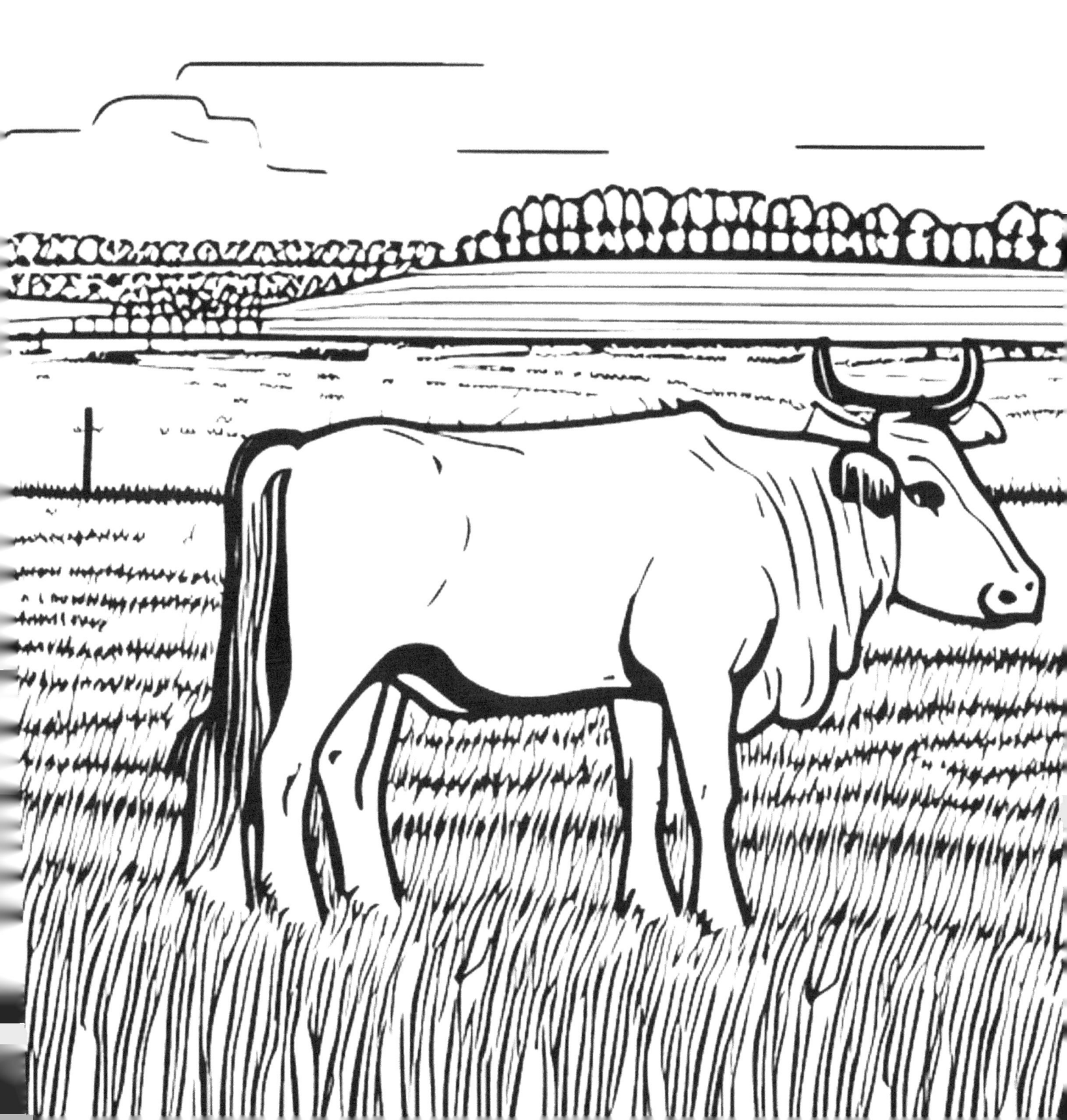

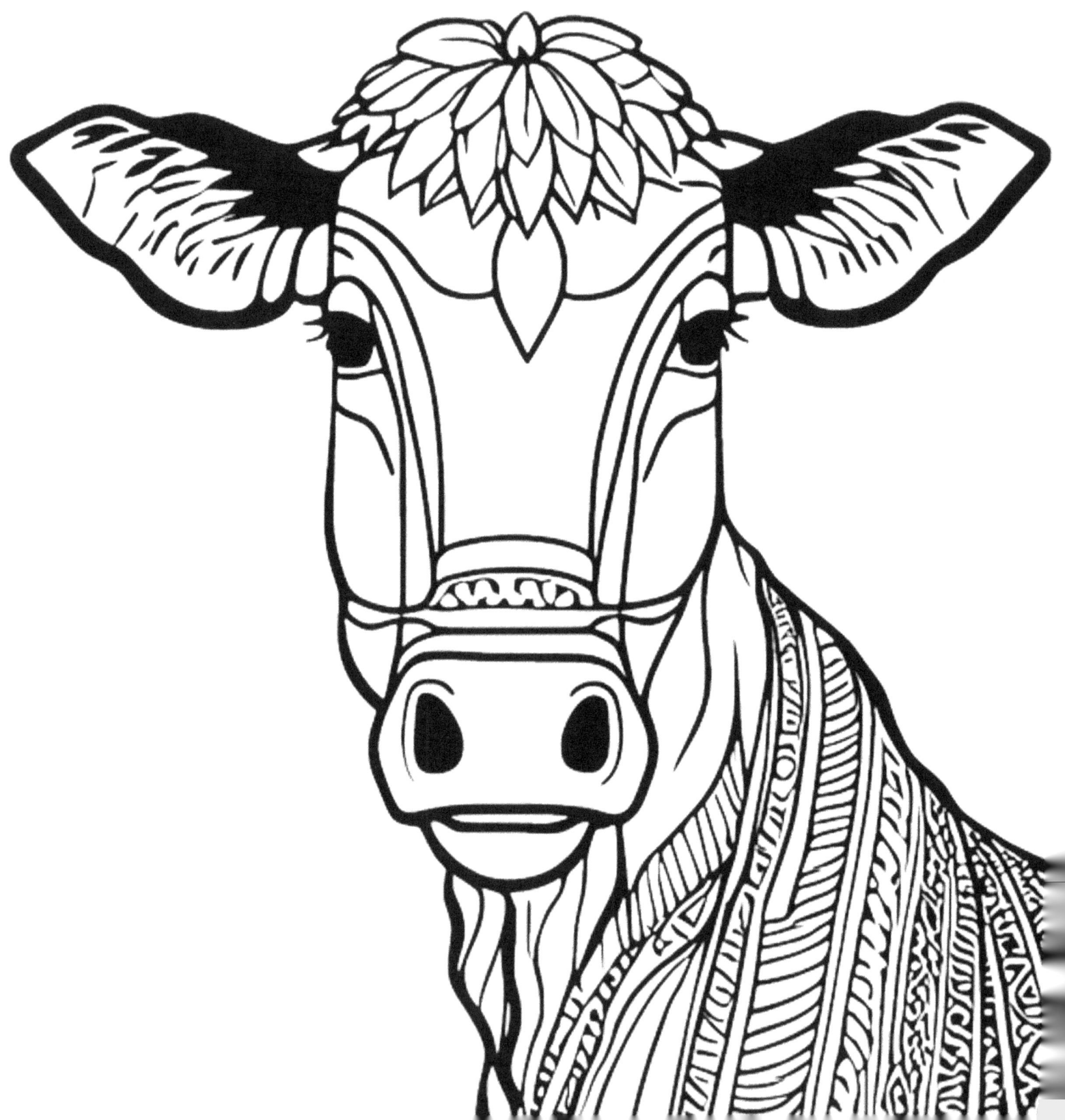

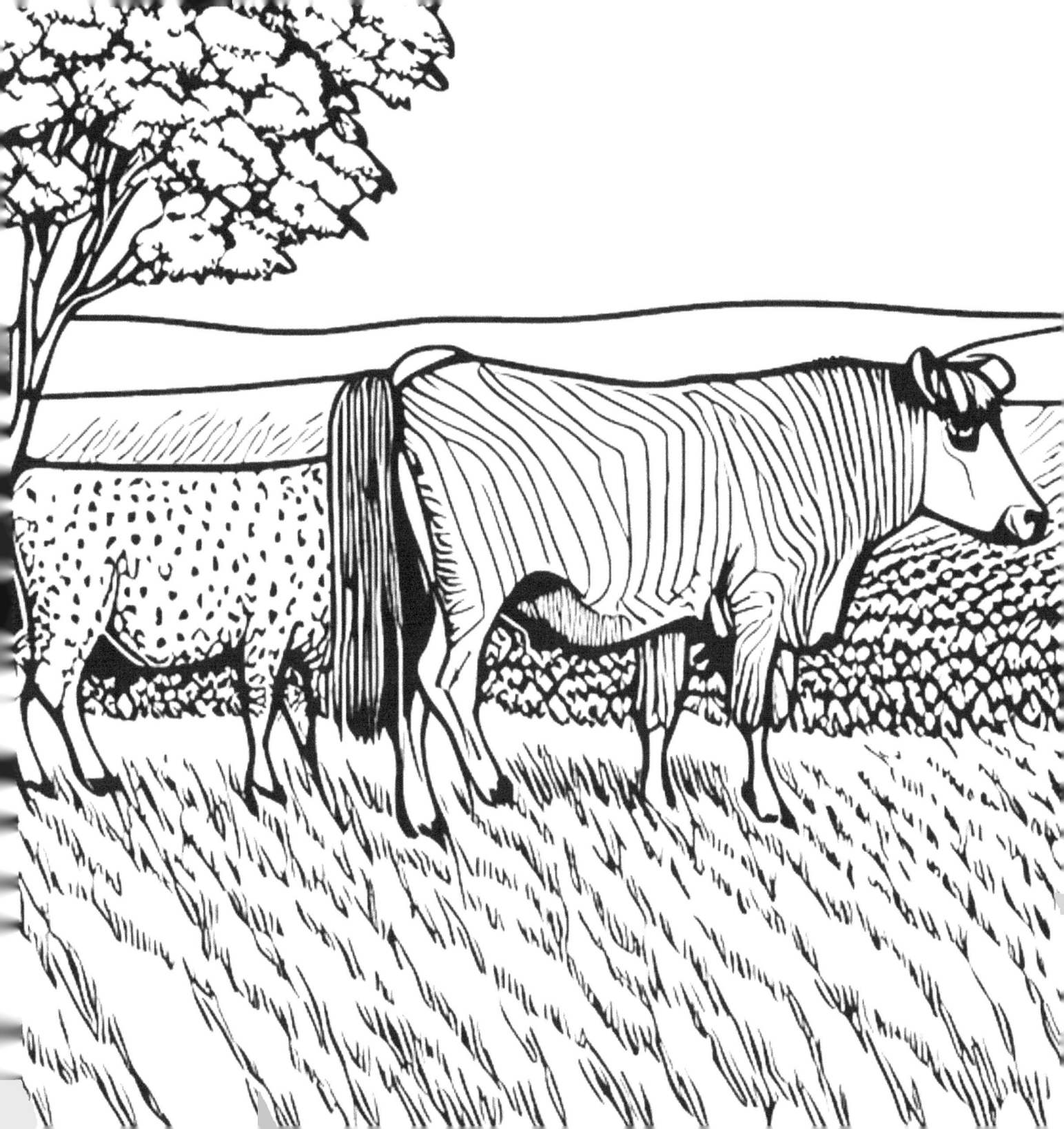

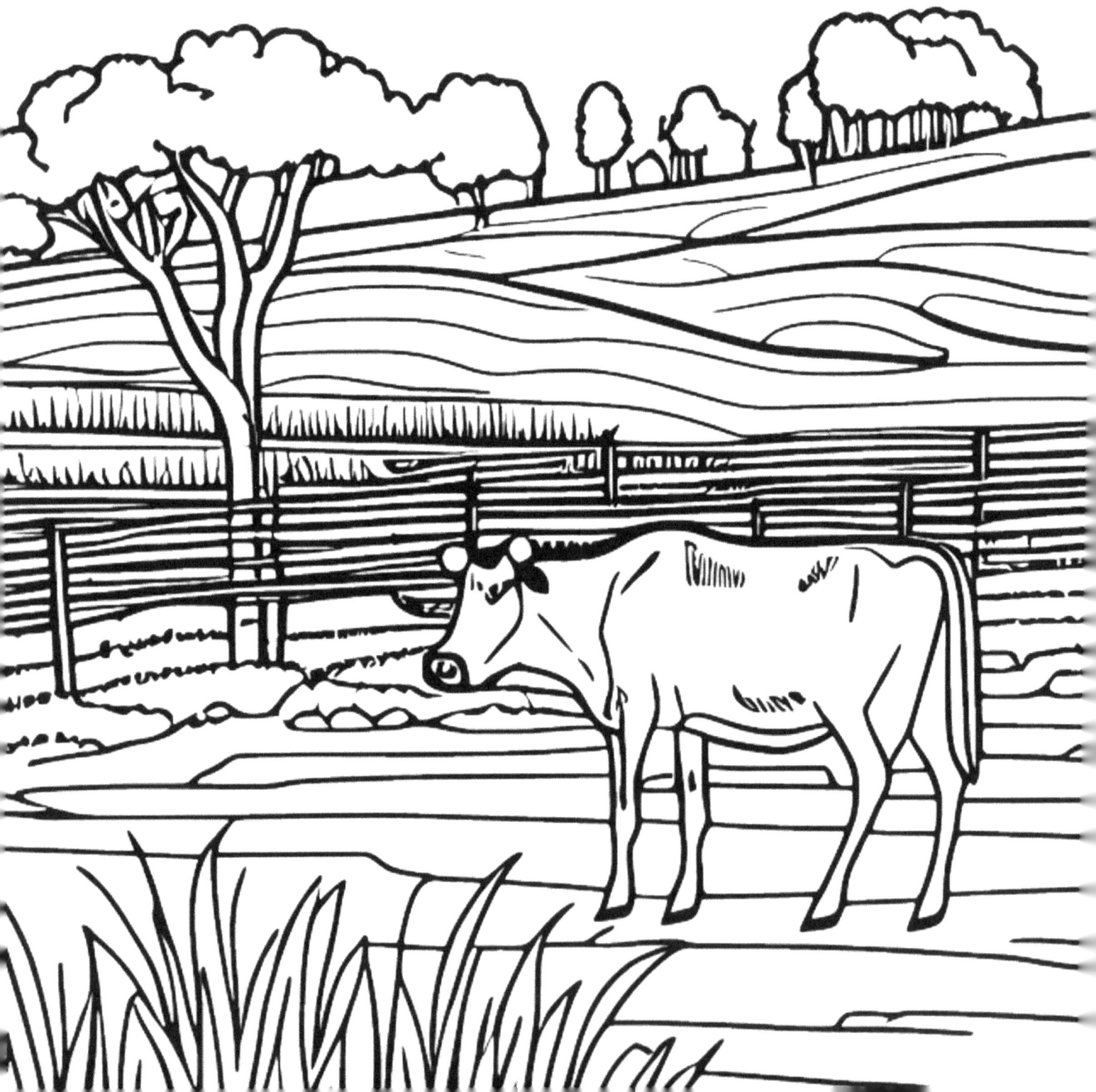

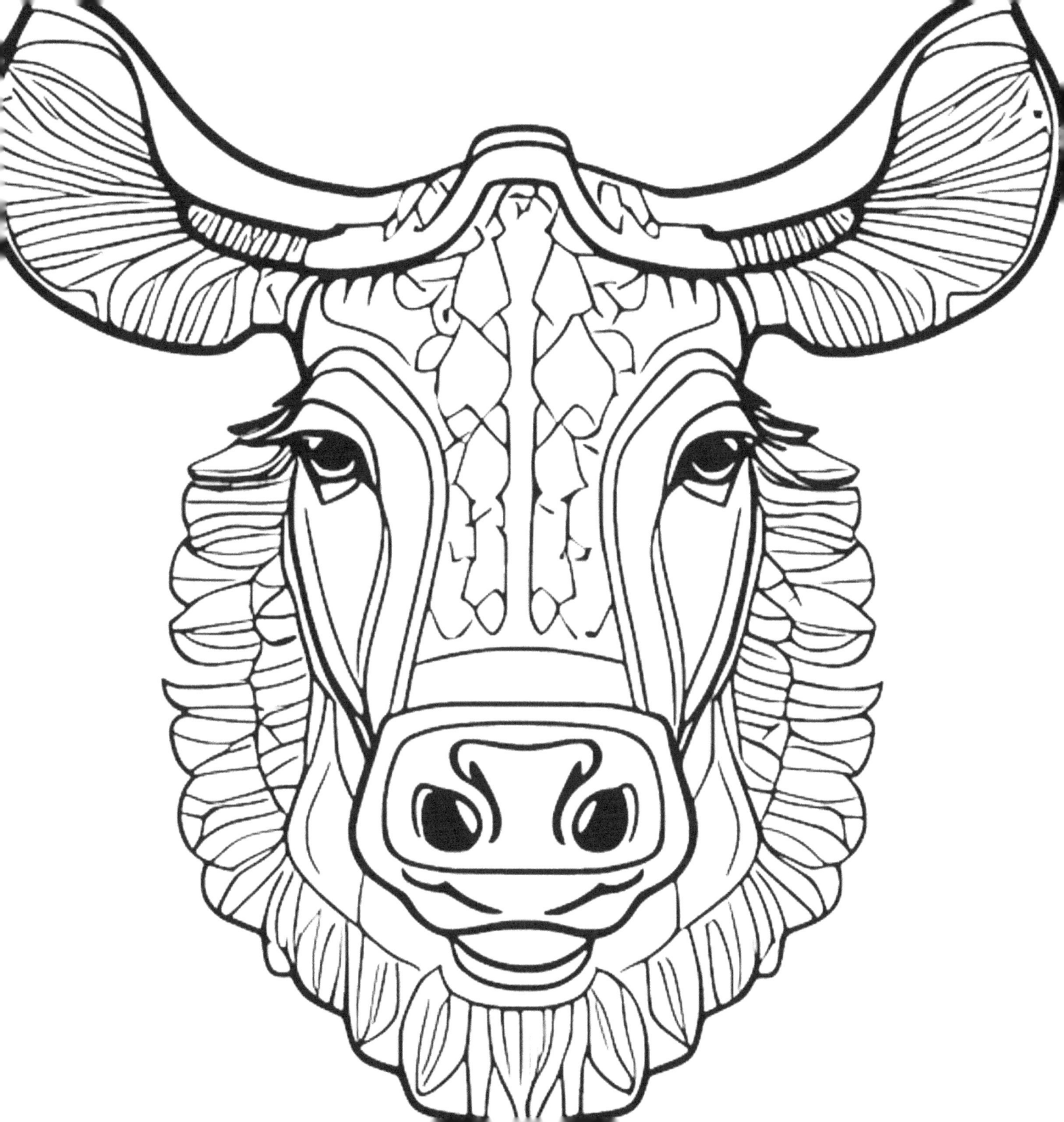

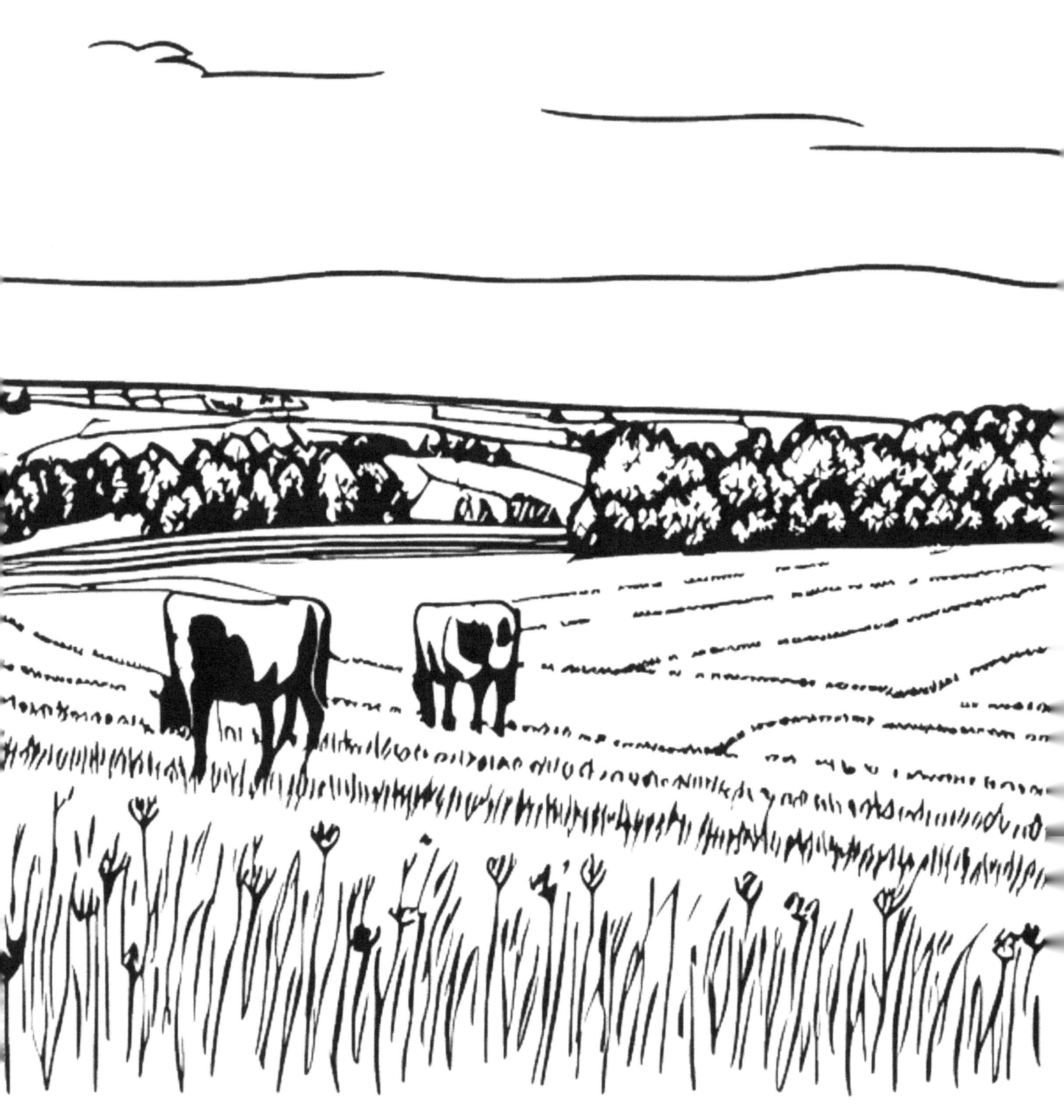

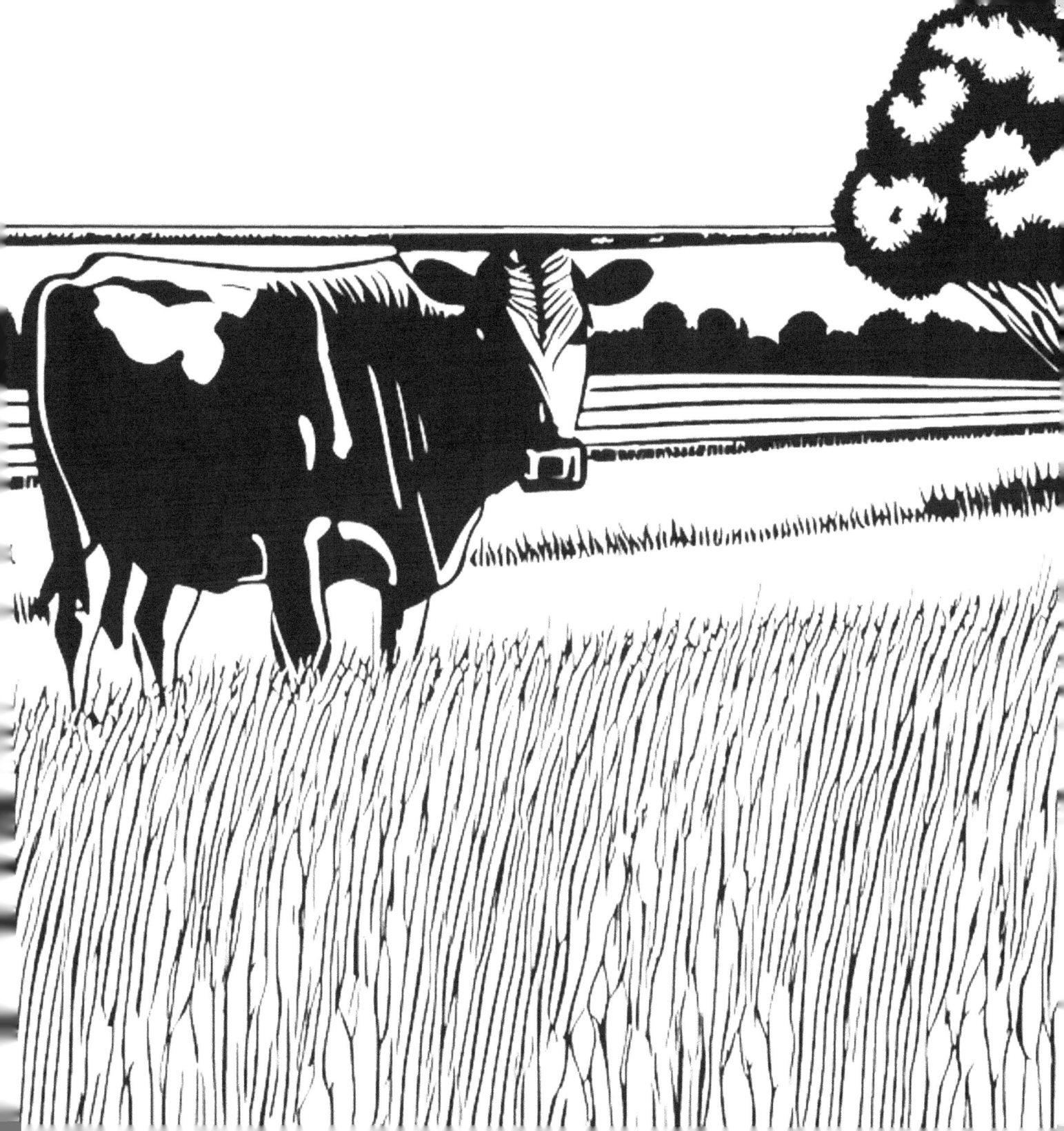

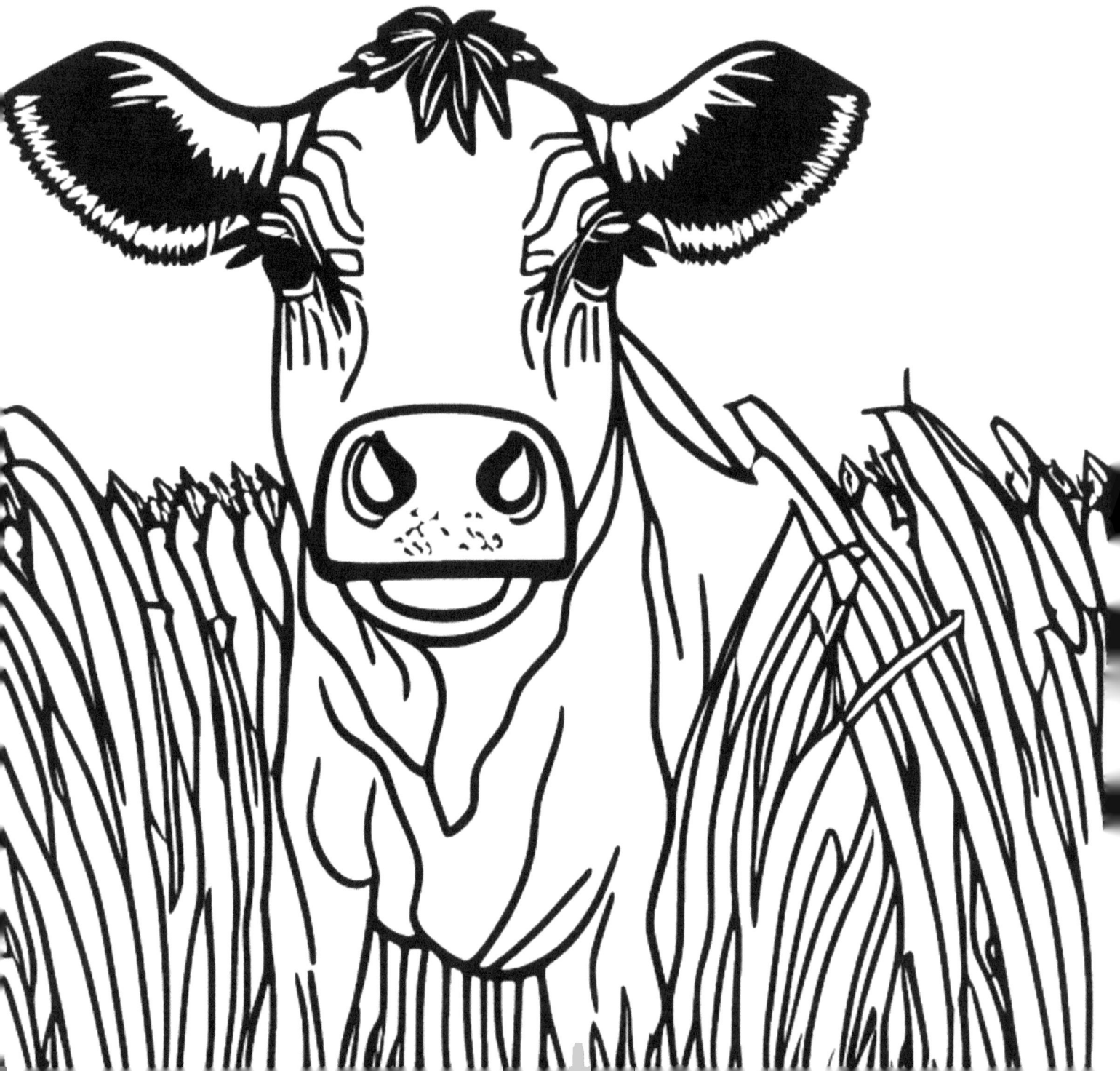

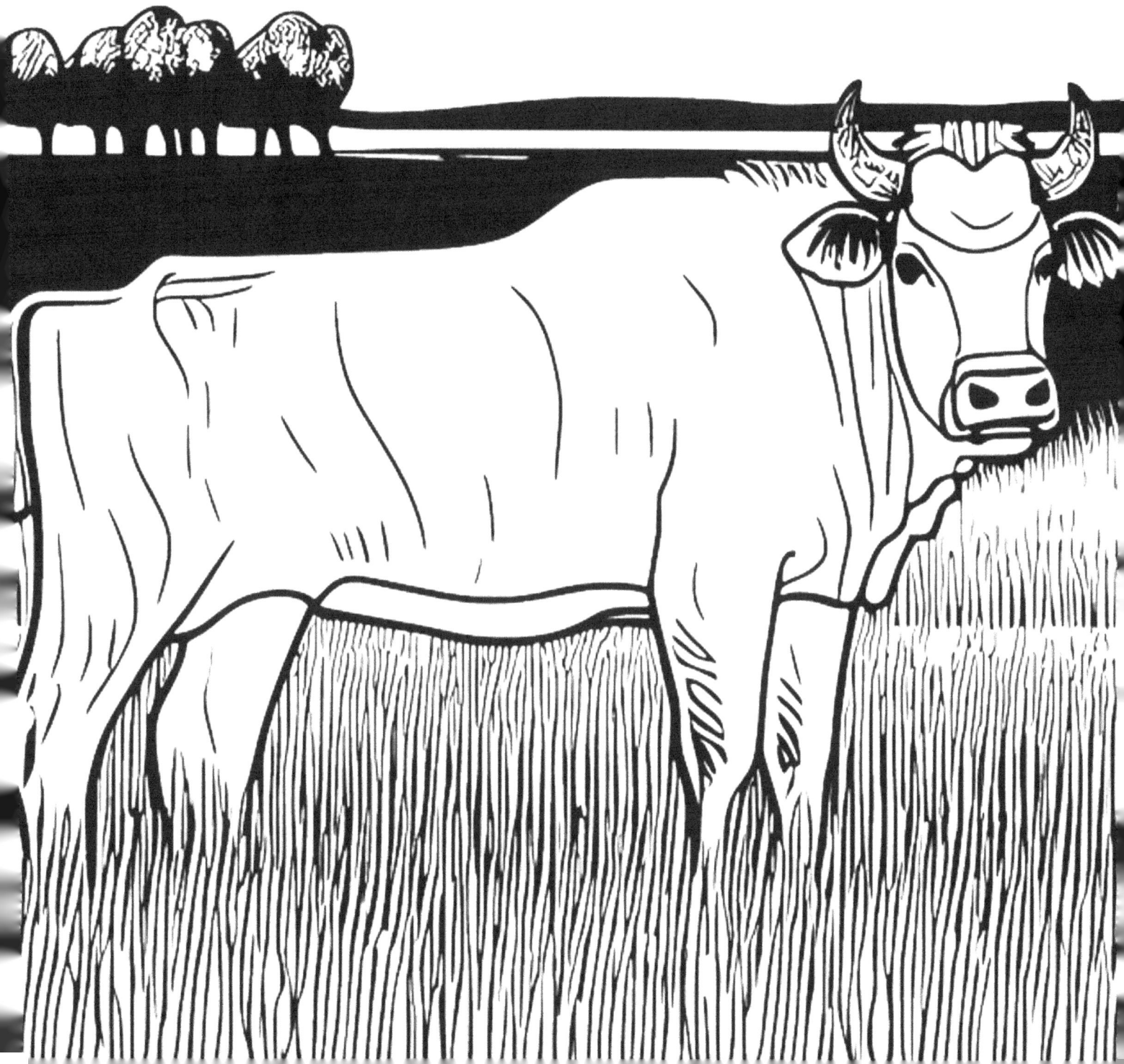

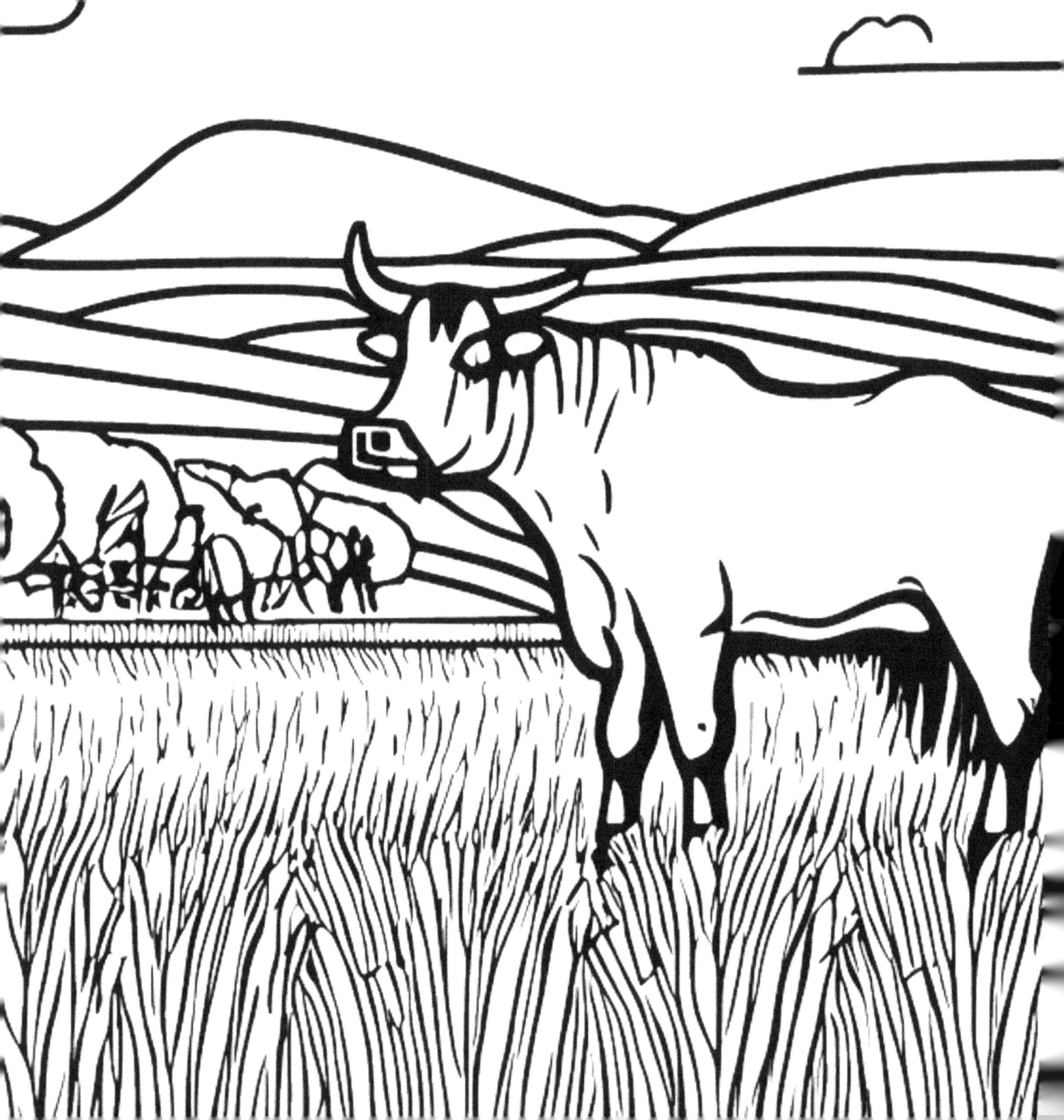

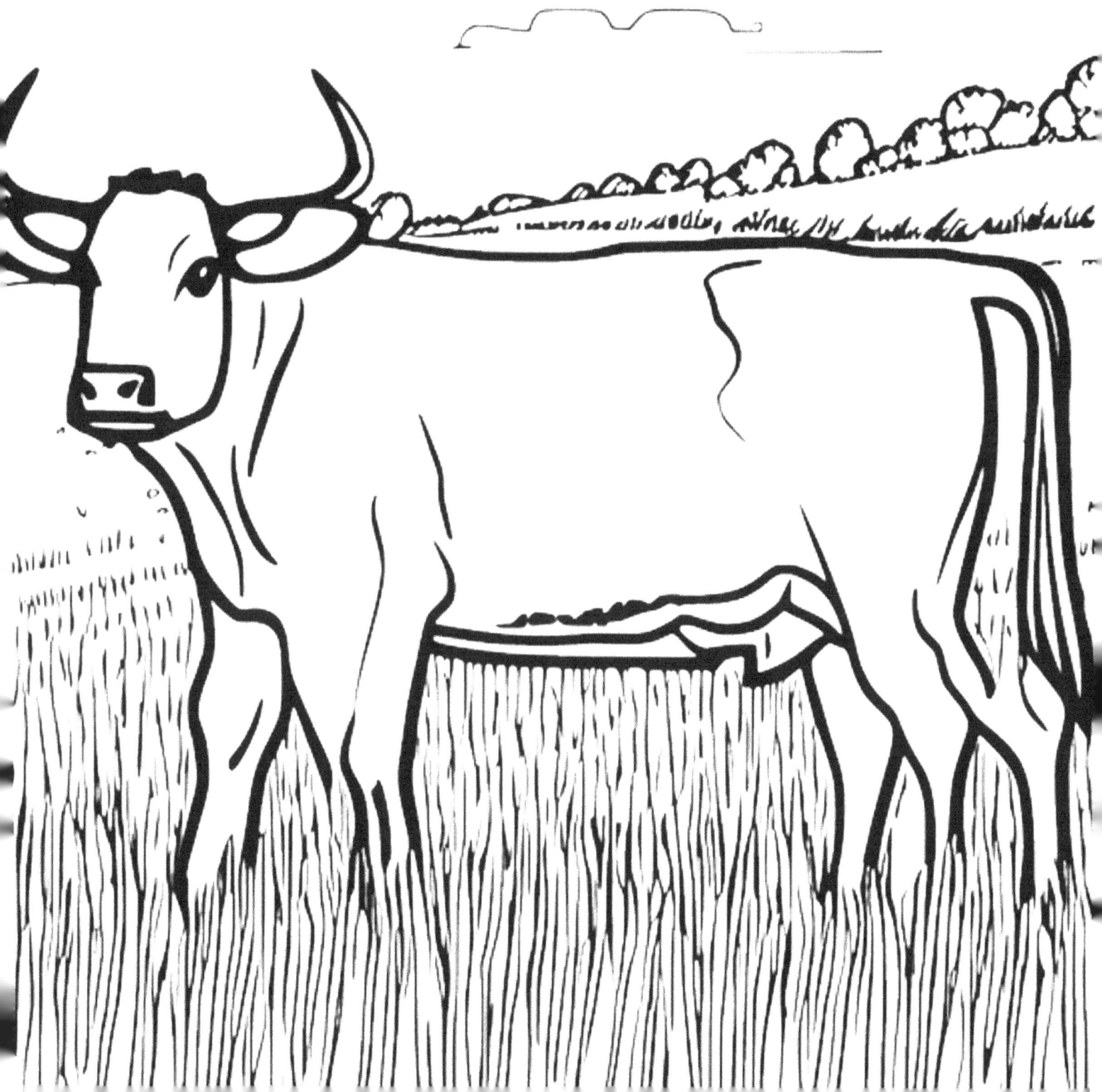

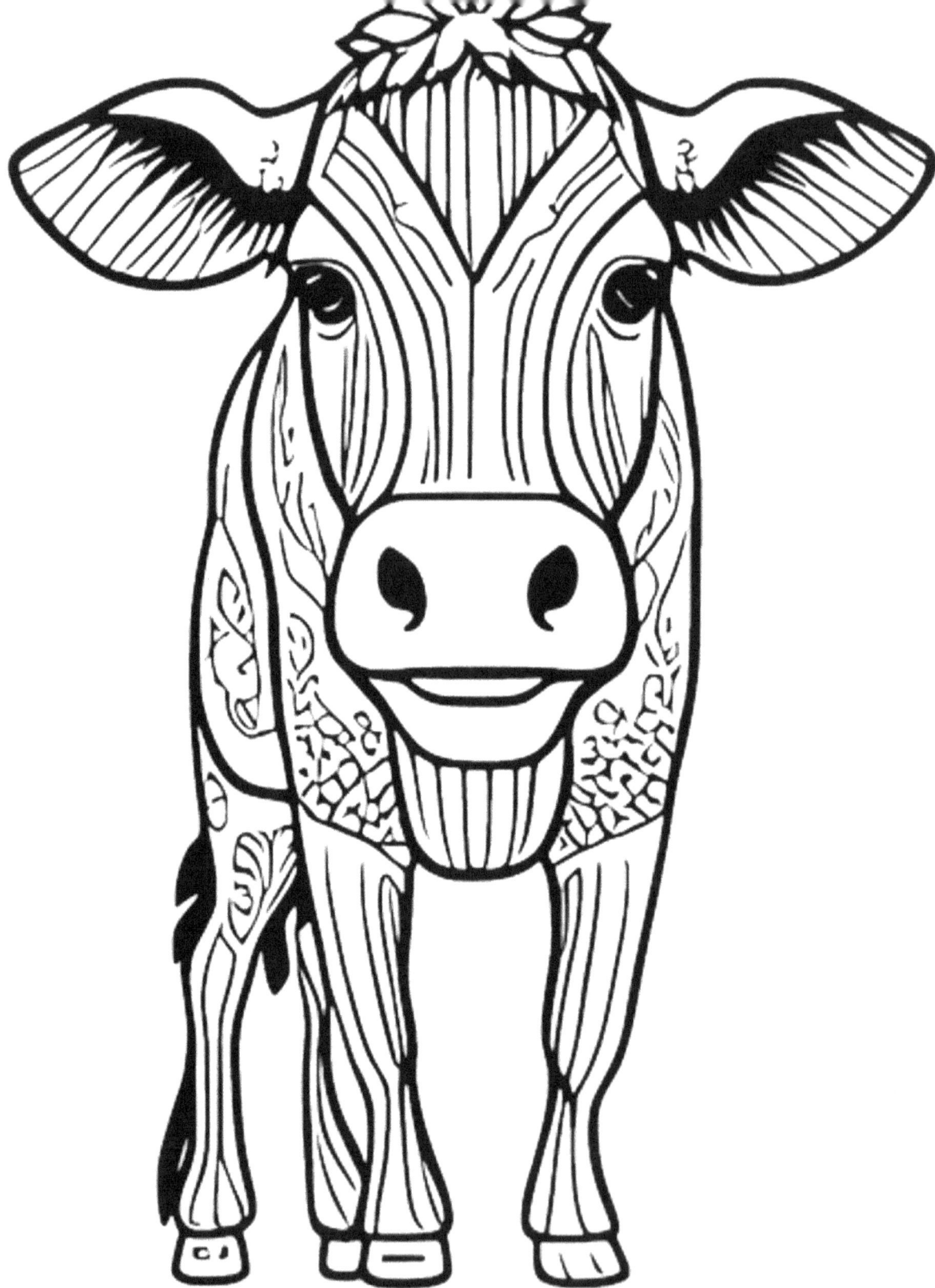

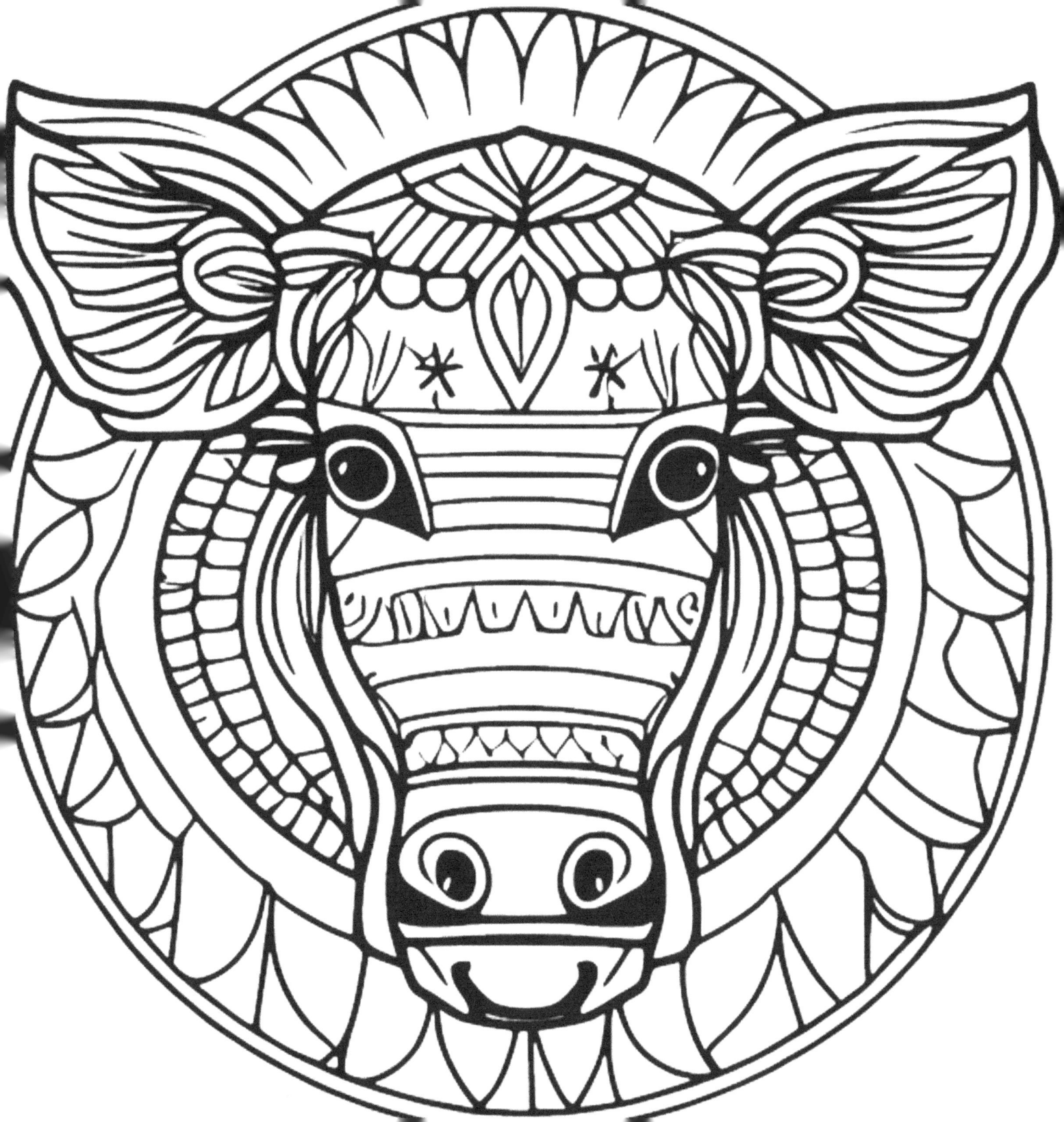

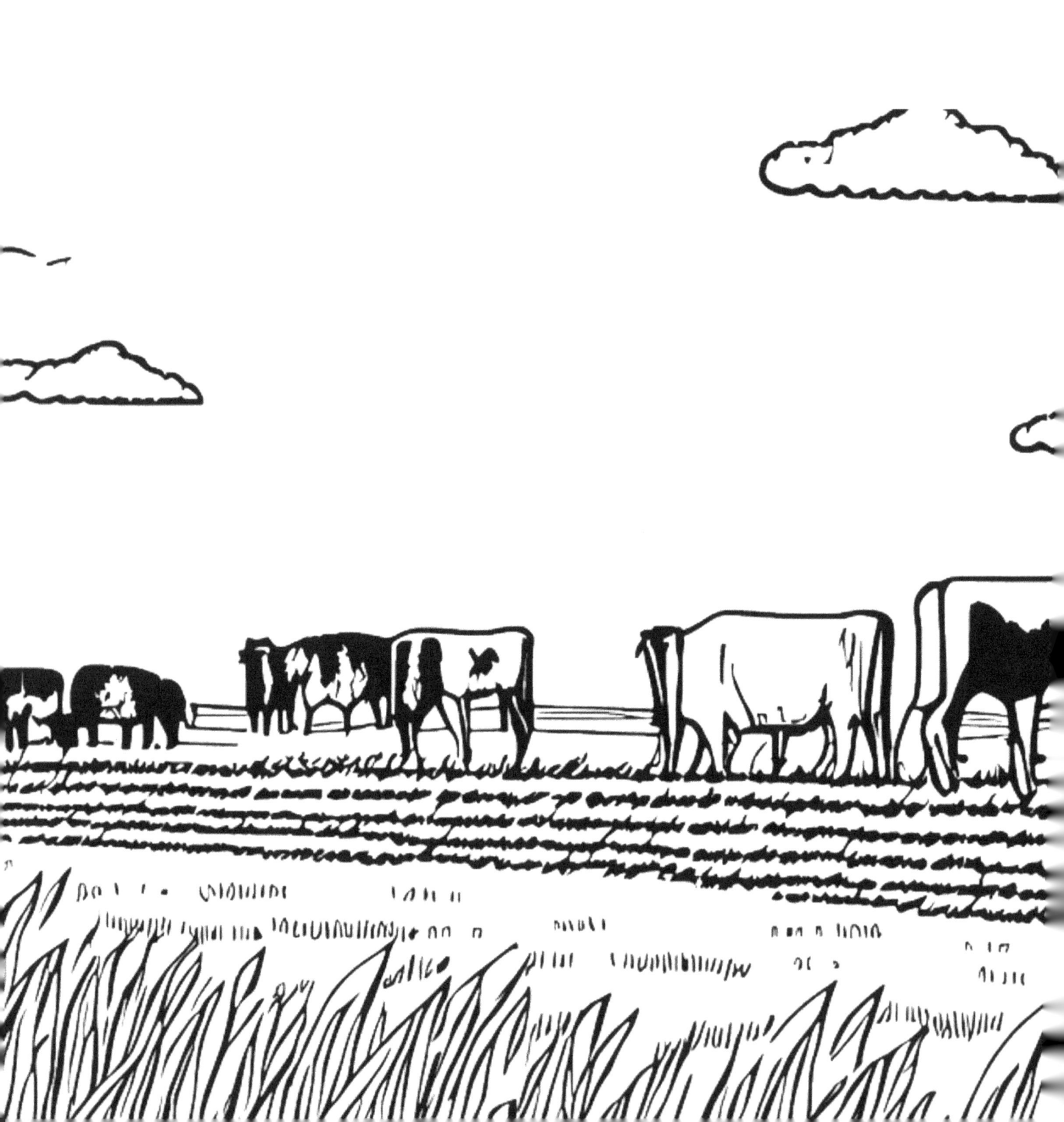

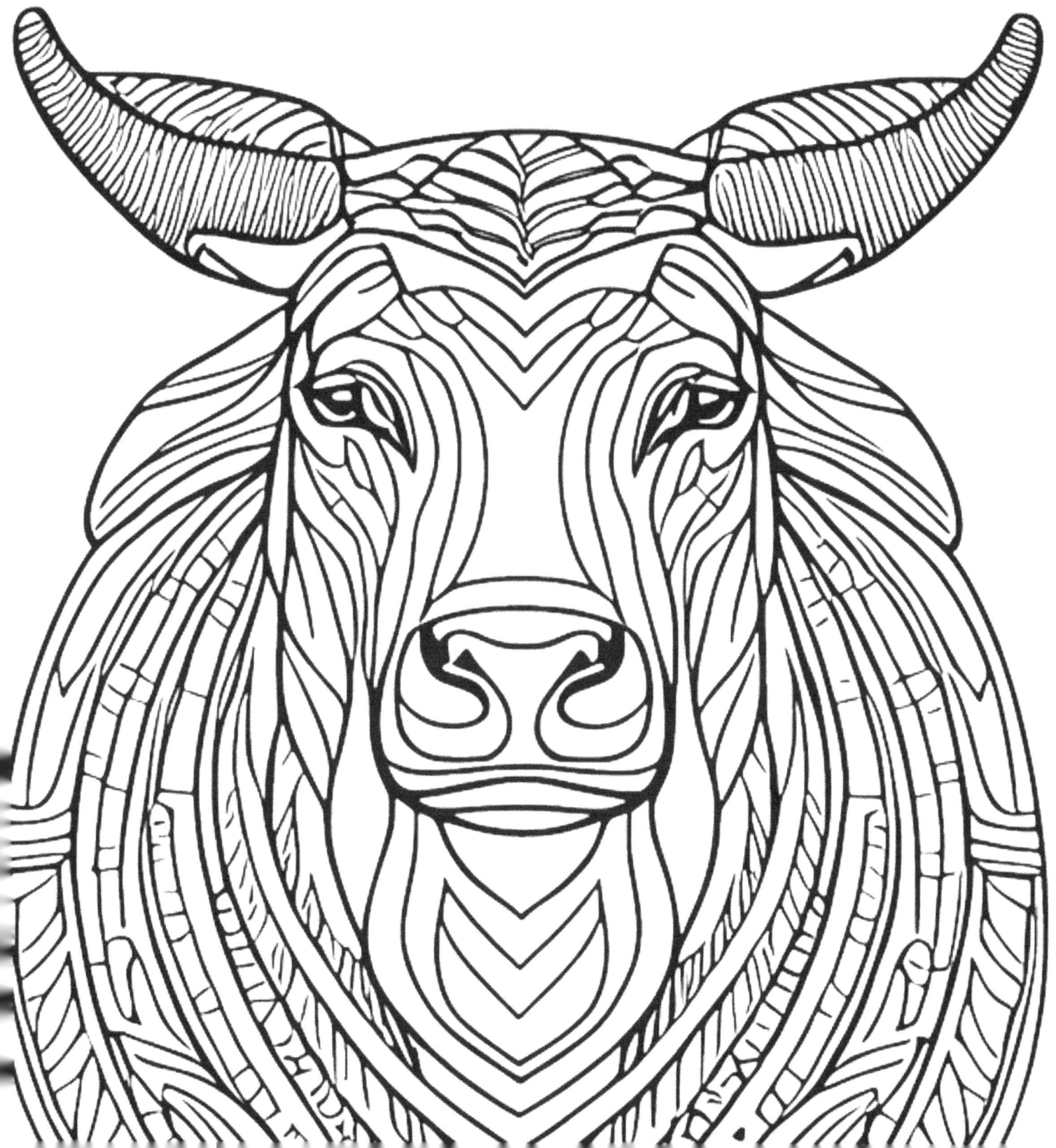

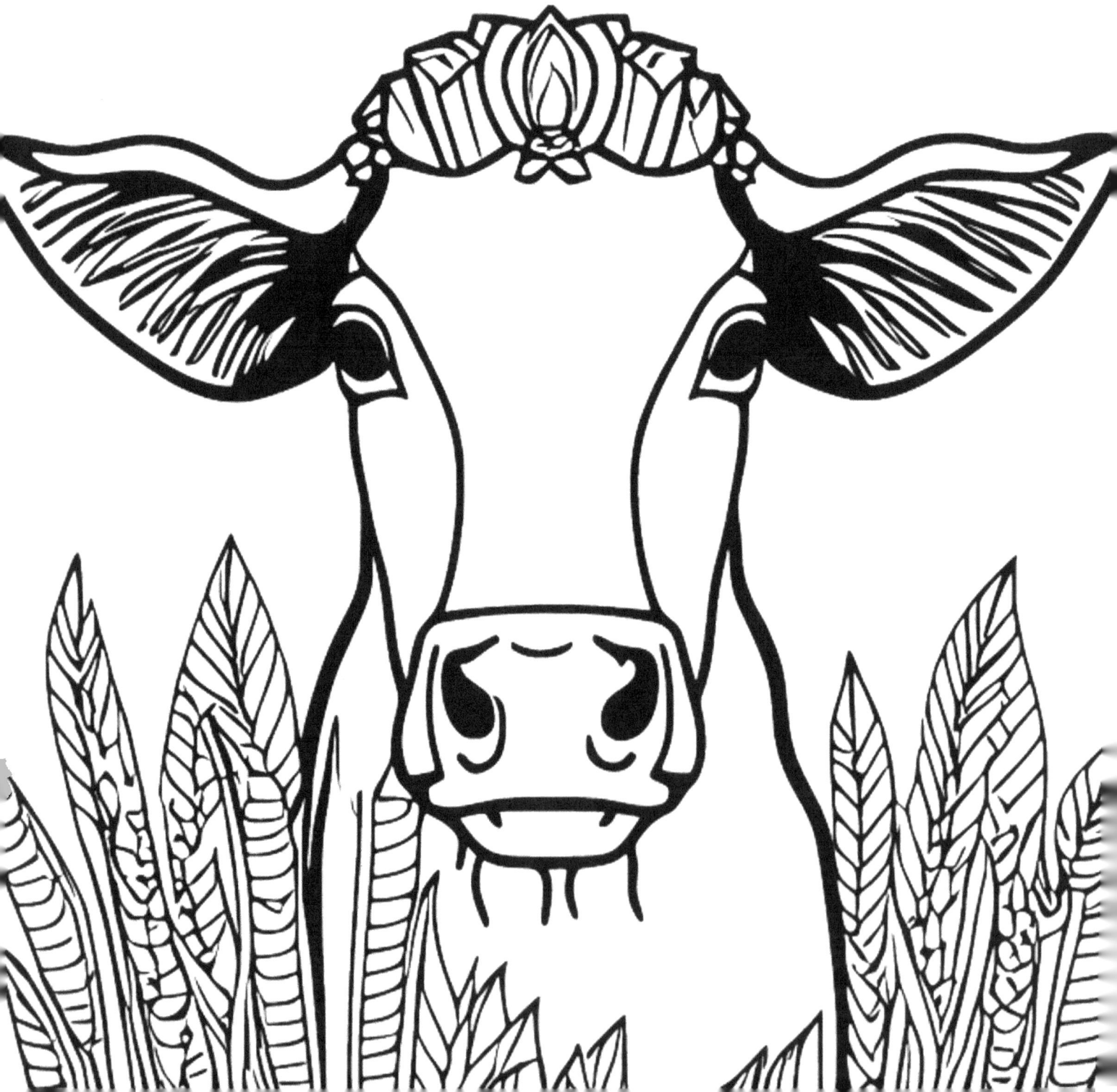

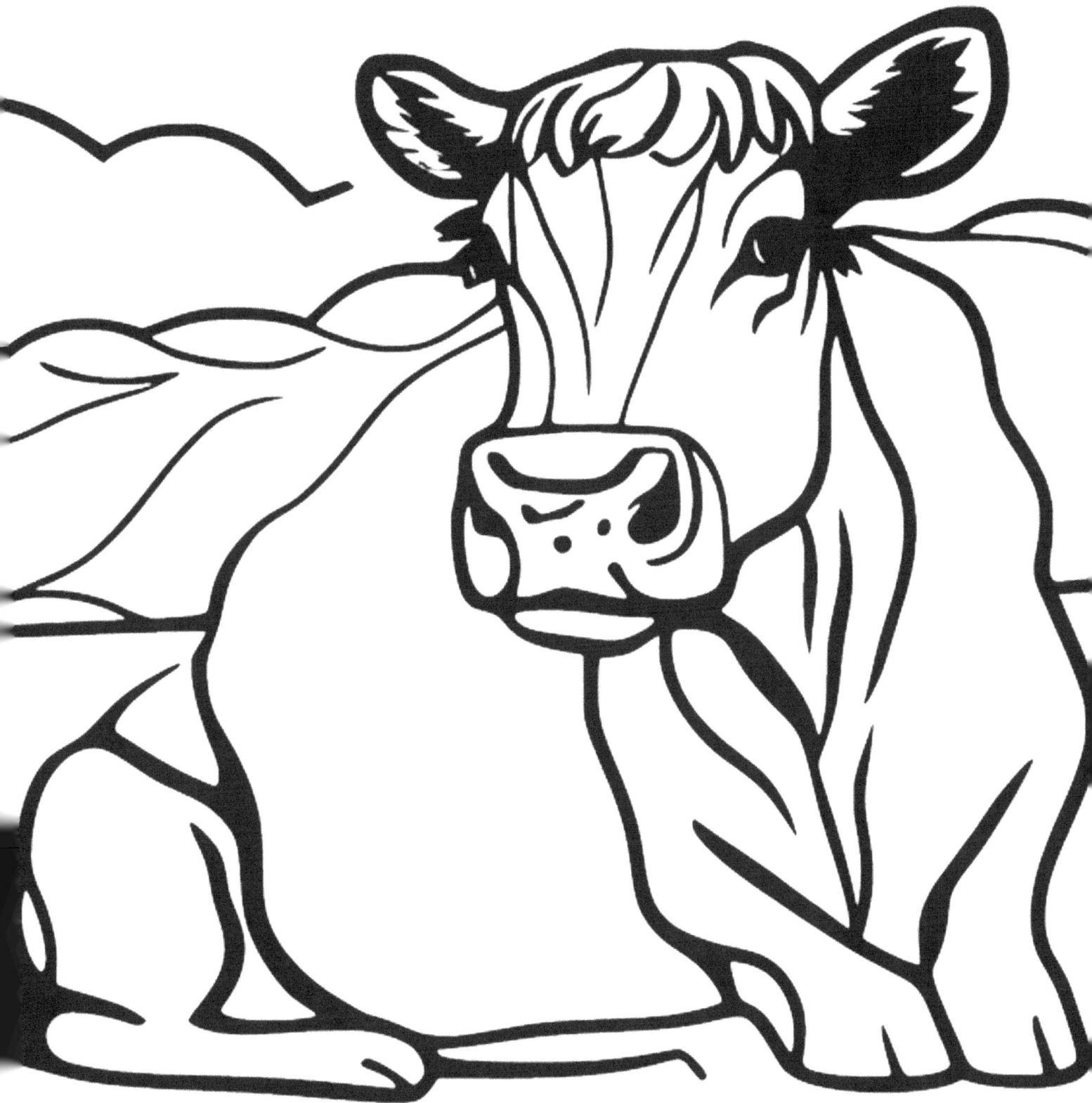

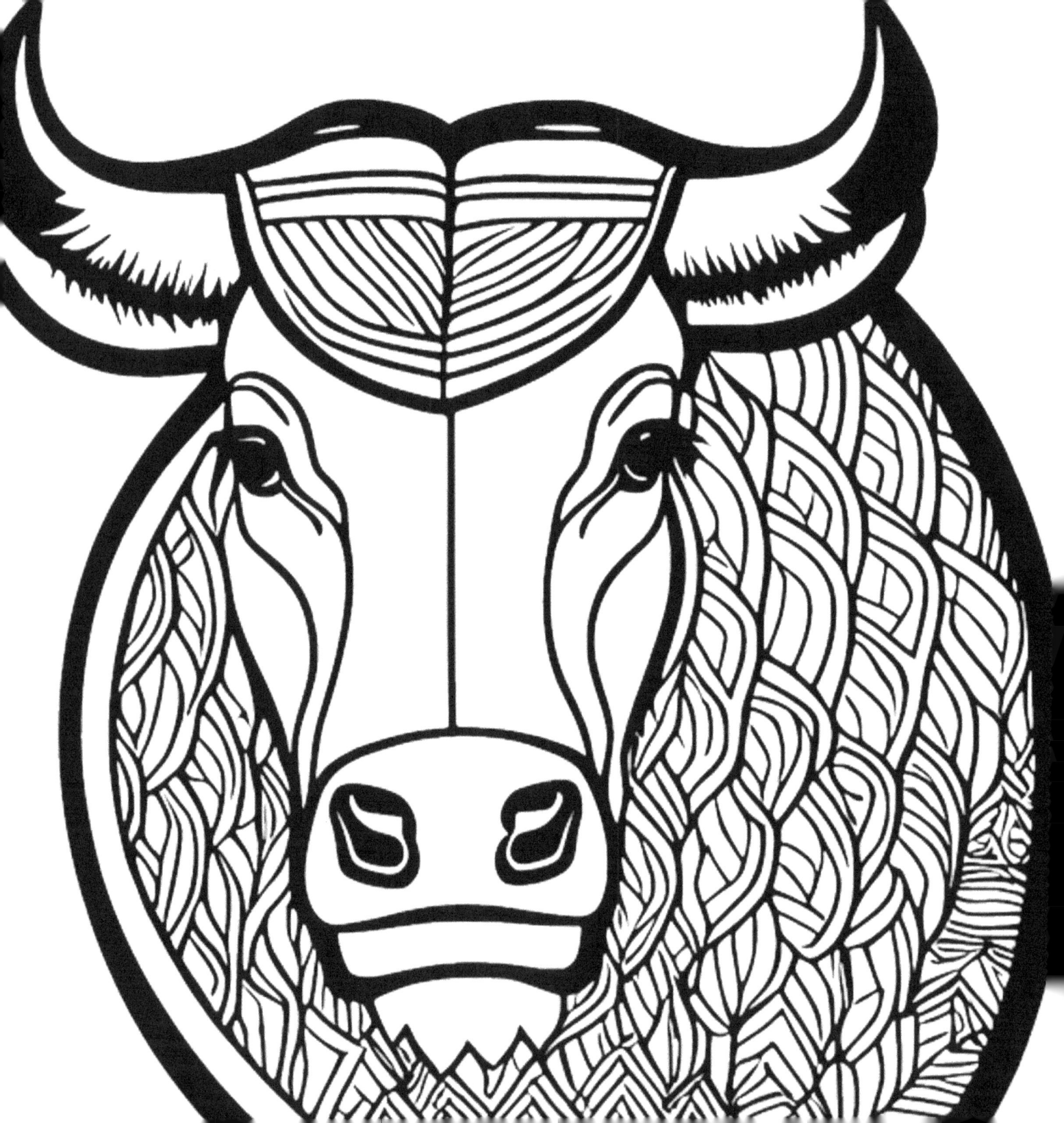

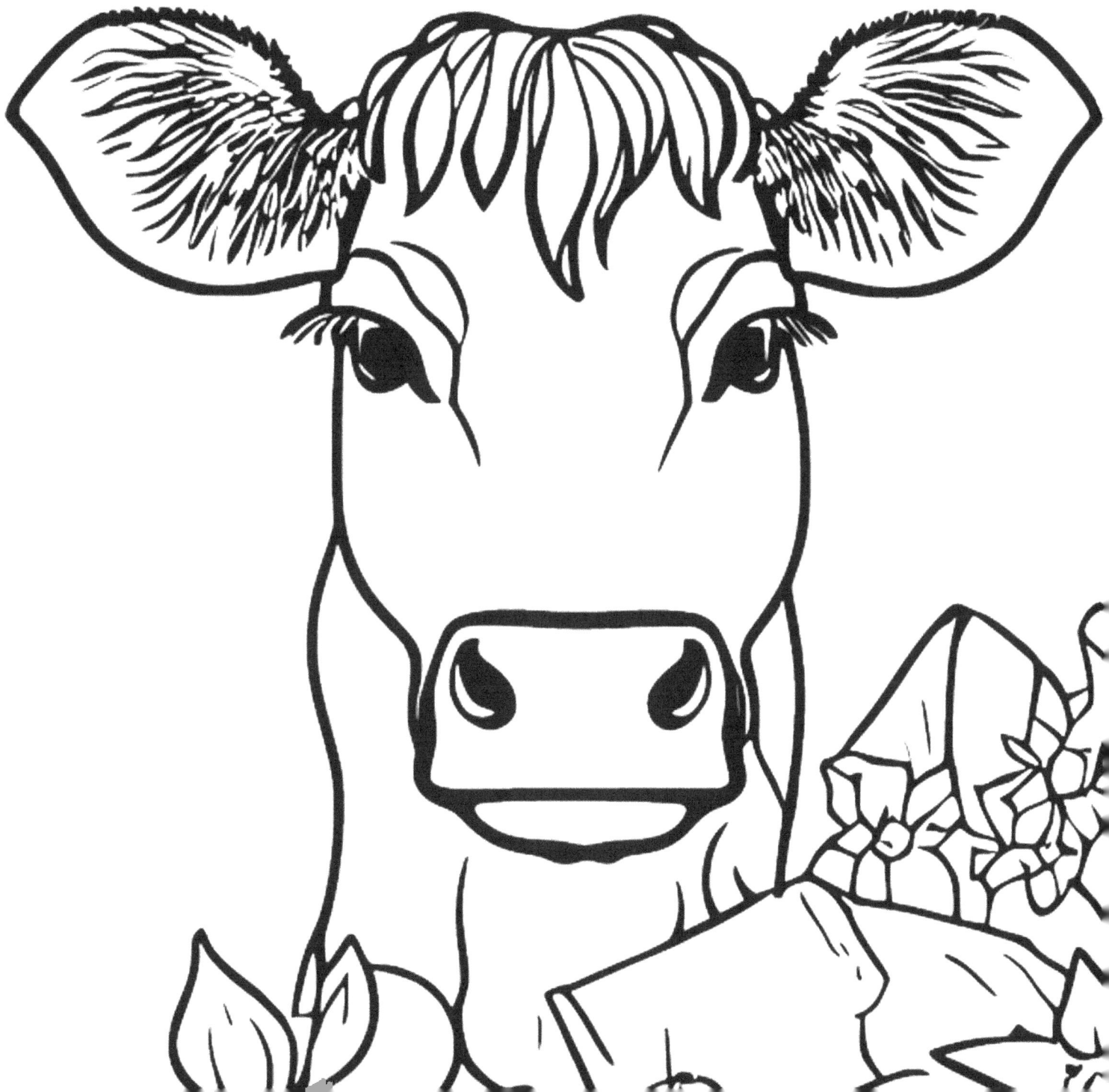

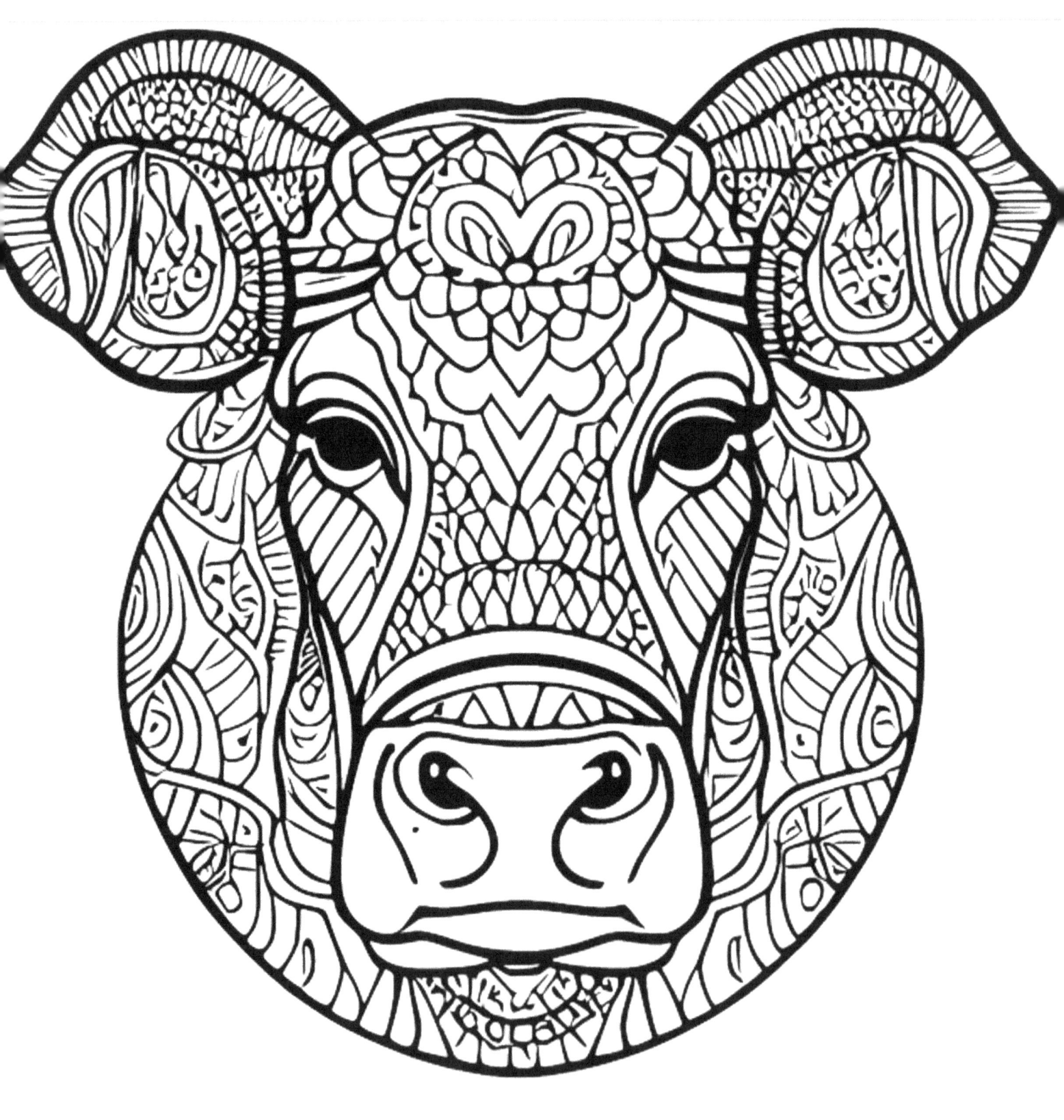

www.ingramcontent.com/pod-product-compliance
Lightning Source LLC
Chambersburg PA
CBHW052346210326
41597CB00037B/6273